Effective Solutions to Pollution Mitigation for Public Welfare

Ahmet Ozan Gezerman
Yildiz Technical University, Turkey

Burcu Didem Corbacioglu
Yildiz Technical University, Turkey

Bhola R. Gurjar
Indian Institute of Technology Roorkee, India

A volume in the Advances in Environmental Engineering and Green Technologies (AEEGT) Book Series

IGI Global
DISSEMINATOR OF KNOWLEDGE

Published in the United States of America by
IGI Global
Engineering Science Reference (an imprint of IGI Global)
701 E. Chocolate Avenue
Hershey PA, USA 17033
Tel: 717-533-8845
Fax: 717-533-8661
E-mail: cust@igi-global.com
Web site: http://www.igi-global.com

Library of Congress Cataloging-in-Publication Data

Names: Gezerman, Ahmet Ozan, 1978- editor. | Corbacioglu, Burcu Didem, 1967- editor. | Gurjar, Bhola R., 1965- editor.
Title: Effective solutions to pollution mitigation for public welfare / Ahmet Ozan Gezerman, Burcu Didem Corbacioglu, and Bhola R. Gurjar, editors.
Description: Hershey, PA : Engineering Science Reference, [2018] | Includes index. | Includes bibliographical references and index.
Identifiers: LCCN 2017028942| ISBN 9781522533795 (hardcover) | ISBN 9781522533801 (ebook)
Subjects: LCSH: Pollution. | Environmental protection. | Environmental policy.
Classification: LCC TD174 .E36 2018 | DDC 363.73/6--dc23 LC record available at https://lccn.loc.gov/2017028942

This book is published in the IGI Global book series Advances in Environmental Engineering and Green Technologies (AEEGT) (ISSN: 2326-9162; eISSN: 2326-9170)

British Cataloguing in Publication Data
A Cataloguing in Publication record for this book is available from the British Library.

Advances in Environmental Engineering and Green Technologies (AEEGT) Book Series

ISSN:2326-9162
EISSN:2326-9170

Editors-in-Chief: Sang-Bing Tsai, University of Electronic Science and Technology of China Zhongshan Institute, China & Ming-Lang Tseng, Lunghwa University of Science and Technology, Taiwan & Yuchi Wang, University of Electronic Science and Technology of China Zhongshan Institute, China

MISSION

Growing awareness and an increased focus on environmental issues such as climate change, energy use, and loss of non-renewable resources have brought about a greater need for research that provides potential solutions to these problems. Research in environmental science and engineering continues to play a vital role in uncovering new opportunities for a "green" future.

The **Advances in Environmental Engineering and Green Technologies (AEEGT)** book series is a mouthpiece for research in all aspects of environmental science, earth science, and green initiatives. This series supports the ongoing research in this field through publishing books that discuss topics within environmental engineering or that deal with the interdisciplinary field of green technologies.

COVERAGE

- Sustainable Communities
- Contaminated Site Remediation
- Radioactive Waste Treatment
- Pollution Management
- Green Transportation
- Industrial Waste Management and Minimization
- Policies Involving Green Technologies and Environmental Engineering
- Green Technology
- Air Quality
- Renewable Energy

IGI Global is currently accepting manuscripts for publication within this series. To submit a proposal for a volume in this series, please contact our Acquisition Editors at Acquisitions@igi-global.com or visit: http://www.igi-global.com/publish/.

The Advances in Environmental Engineering and Green Technologies (AEEGT) Book Series (ISSN 2326-9162) is published by IGI Global, 701 E. Chocolate Avenue, Hershey, PA 17033-1240, USA, www.igi-global.com. This series is composed of titles available for purchase individually; each title is edited to be contextually exclusive from any other title within the series. For pricing and ordering information please visit http://www.igi-global.com/book-series/advances-environmental-engineering-green-technologies/73679. Postmaster: Send all address changes to above address.

Titles in this Series

For a list of additional titles in this series, please visit:
https://www.igi-global.com/book-series/advances-environmental-engineering-green-technologies/73679

Green Production Strategies for Sustainability
Sang-Binge Tsai (University of Electronic Science and Technology of China (Zhongshan Institute), China & Civil Aviation University of China, China) Bin Liu (Shanghai Maritime University, China) and Yongian Li (Nankai University,China)
Engineering Science Reference • ©2018 • 325pp • H/C (ISBN: 9781522535379) • US $235.00

Innovative Strategies and Frameworks in Climate Change Adaptation Emerging Research ...
Alexander G. Flor (University of the Philippines Open University, Philippines) and Benjamina Gonzalez Flor (University of the Philippines, Philippines)
Engineering Science Reference • ©2018 • 165pp • H/C (ISBN: 9781522527671) • US $165.00

Utilizing Innovative Technologies to Address the Public Health Impact of Climate Change ...
Debra Weiss-Randall (Florida Atlantic University, USA)
Engineering Science Reference • ©2018 • 295pp • H/C (ISBN: 9781522534143) • US $185.00

Economical and Technical Considerations for Solar Tracking Methodologies...
S. Soulayman (Higher Institute for Applied Sciences and Technology, Syria)
Engineering Science Reference • ©2018 • 647pp • H/C (ISBN: 9781522529507) • US $245.00

Hydrology and Best Practices for Managing Water Resources in Arid and Semi-Arid Lands
Christopher Misati Ondieki (Kenyatta University, Kenya) and Johnson U. Kitheka (South Eastern Kenya University, Kenya)
Engineering Science Reference • ©2018 • 266pp • H/C (ISBN: 9781522527190) • US $205.00

Computational Techniques for Modeling Atmospheric Processes
Vitaliy Prusov (University of Kyiv, Ukraine) and Anatoliy Doroshenko (National Academy of Sciences, Ukraine)
Information Science Reference • ©2018 • 460pp • H/C (ISBN: 9781522526360) • US $205.00

For an entire list of titles in this series, please visit:
https://www.igi-global.com/book-series/advances-environmental-engineering-green-technologies/73679

IGI Global
DISSEMINATOR OF KNOWLEDGE

701 East Chocolate Avenue, Hershey, PA 17033, USA
Tel: 717-533-8845 x100 • Fax: 717-533-8661
E-Mail: cust@igi-global.com • www.igi-global.com

Table of Contents

Detailed Table of Contents

This chapter presents the overview of pollution; the issues of soil pollution, water pollution, and air pollution; the aspects of renewable energy; energy security and energy imports; and renewable energy policy and renewable energy policy instruments. Pollution is one of the most important environmental, social, and health issues in the world. Pollution creates many diseases and causes death of many people across the globe. The environmental damage caused by pollution can reach catastrophic proportions and destroy entire ecosystems leading to the death of many species and a big biodiversity loss. Renewable energy is a critical part of reducing global carbon emissions and the pace of investment has greatly increased as the cost of technologies fall and efficiency continues to rise. Renewable energy offers a wide variety of different options to choose from as countries can choose between sun, wind, biomass, geothermal energy, and water resources.

Exposure analysis is the receptor-oriented approach of the pollution-level measurement. In this chapter, a detailed discussion is provided of the fundamentals of exposure analysis, methods of measurement, basics of models used for the prediction of pollution concentration indoors and outdoors, and a brief discussion about the health impact of selected pollutants. A detail of fundamental of indoor air quality (IAQ) models like mass balance and CFD models is discussed. Also, basic structures of community multiscale air quality model (CMAQ) and AIRMOD ambient air dispersion models are described. It is observed that measurement of pollution exposure by direct method requires more time and effort as compared with the integrated exposure and stationary measurement. AIRMODE is steady state model and based upon the Gaussian dispersion model. CMAQ is capable of simulating the pollution level for the range of geographic scale for multiple pollutants.

Carbon dioxide emissions cause global warming, and greenhouse gases and climate change are very serious problems. Mesoporous silica material SBA-15 has been preferred mostly as an ideal adsorbent for CO2 due to its excellent properties such as high surface areas and pore volumes, larger pore diameter, and thicker silica wall. In the literature studies, SBA-15 has been modified by different functional groups and the effects of modification methods on the CO2 adsorption have been investigated. Modified SBA-15 adsorbents showed high CO2 adsorption capacity. The aim of this chapter is to review the use of modified-SBA-15 mesoporous silica materials as adsorbent for CO2 capture.

Today, there are many studies about the problems that may be faced in the context of World Water Day. In this chapter, the structure, pollution, quality grading, and human health effects of water; possible pollution prevention measures; and water safety are discussed. It is expected that the world population, which is about 7 billion currently, will rise to 9 billion by 2050. Water consumption is expected to increase at a higher rate, which is a major problem for the environment. By 2025, it has been estimated that two-thirds of the world's population will deal with water shortage. The world is not as rich in water as once thought and, hence, is at high risk for water shortage. For these reasons, we must all fulfill our responsibility to leave a habitable world to future generations.

The enhancements in the socio-economic status of many people has come from the expansion of agricultural and industrial production. But, some of the activities associated with this expansion have adversely affected water quality. This leads to a negative impact on public health, eminence of life, and environment. This chapter sets out to explain the various factors that lead to water contamination and different mitigation techniques to manage them. We need this knowledge so as to develop suitable solutions for a broad range of environmental problems.

Foreword

Environmental pollution of the air, water, soil, and so on is caused by human activities. To prevent increasing environmental pollution, many local and worldwide studies are being conducted with the contributions of various sciences. These studies deal with environmental pollution on different dimensions and recommend solutions.

To adequately meet the needs of an increasing population, intense industrialization has become a necessity. However, problems have emerged in terms of the environmental pollution due to enterprises near settlement centers. The problems include harmful waste, a lack of proper storage, lack of lime deposits in pits, lack of projected warehouses with sufficient operational capacity, and the environmental pollution resulting from the odors and image pollution caused by these adverse conditions.

Therefore, legal and technical precautions and storage and forecasting criteria should be examined to prevent the waste generated by enterprises from creating adverse environmental conditions.

The storage of waste products from industrial producers results in water pollution. Unlike urban pollution sources, those generated by enterprises are not point sources, but span a wider area. This makes it more difficult to determine the dimensions of water pollution caused by these sources. Industrial waste, which is classified as a scattered source of pollution, causes surface water or infiltration to reach the lower layers of the soil and groundwater, disrupting the quality of water resources and rendering them unusable. Organic waste causes water pollution, because of its high biochemical oxygen requirement. In addition, organic waste is a potential source of pathogenic contamination as well as nitrogen and phosphorus in the water. The most adverse impact of industrial production on the environment is that it is a source of disease agents. These factors are propagated in the environment through direct and indirect pulses. Furthermore, the pits in which industrial waste is stored are a major source of disease for humans and animals. Note that some illnesses caused by this waste can survive in nature for around one week to three years, indicating that environmental pollution is a long-term problem.

In terms of the quantities and effects of organic pollution, the first factor to be investigated is emission gases. Emission gases, defined as organic waste material, cause environmental pollution internally and externally. This is directly related to the diffusion of domestic pollution. Incorrect transportation of waste in the outdoor environment and improper storage affect and increase first-degree pollution. To avoid adverse environmental conditions and an unhealthy living environment, precautions must be taken and the legal and technical standards as well as storage and forecasting criteria determined. Various forms of waste that cause various levels of damage are produced in industrial facilities. Awareness of their harmful effects on the environment and to animals and human beings is crucial. Furthermore, the effects should be minimized through appropriate projection and applications. Otherwise, industrial plants will continue to operate as sources of pollutants.

The internal environment of industrial facilities includes the temperature, moisture ventilation, and various gases and dust. The industrial products created through the processes applied in these facilities constitute the external environment. Outside the industrial plant, increasing environmental pollution is the result of applied processes. Therefore, the removal of harmful waste that affects human health inside and outside the industrial facility should be kept below the tolerance values that do not harm the environment.

In this book, the environmental impacts of industrial production and the current situation of related facilities are evaluated by considering principles pertaining to settlement planning. In addition, measures and techniques to remove waste from the environment without creating environmental pollution are discussed.

Güldem Üstün
Istanbul Technical University, Turkey

Preface

The concept of "clean production," under consideration since the 1990s on the world agenda, has gained a new dimension with the concept of "green growth." This new dimension has brought about different applications in terms of markets and policy makers. The concept of "clean production" arose in 1999 from scientific-technology-industry discussions platforms and clean production-clean products-eco-friendly technology working group industry sector reports. A clean production center at the proposed national scale has not yet been established. The concept of "clean production" is not sufficiently known and implemented, except for the dimension of energy efficiency. Like most other developing countries, the concept of growth takes priority over that of the environment. Furthermore, adequate capacity building for the environment has not yet been achieved.

CLEAN PRODUCTION

Given the profitability, productivity, competitiveness, and growth objectives of the private sector and the alignment of governments to socio-economic development, growth, productivity, and strategic planning, the private sector is an inseparable part of sustainability and growth.

The use of certain deterrent measures and/or supporting mechanisms of macroeconomic environmental policies is greatly important, especially for developing countries. This is to ensure that pollution limits align with international standards, while taking into account differences in sectors and pollution coefficients.

The clean production approach differs from existing pollution prevention approaches. In existing approaches, the production and design phases are considered as invariable factors, and pollution is seen as an inevitable result. After pollution has occurred, the goal is then to try and solve the problem. Therefore, these approaches focus on better defining pollution and improving waste treatment and disposal, thus often adding significant cost. In clean production, pollution and waste are seen as consequences of inefficiency, in design, in resource utilization, and in production processes. The aim here is providing necessary developments at these stages and therefore providing concurrent economic benefits.

Clean production strategies aim to reduce risks to human health and the environment. They aim for rational inventory control, green logistics for auxiliary steps of production processes, reductions in energy consumption and raw material use for production processes, reductions in toxic substances, and little or no exogenous impacts for services. Environmentally sensitive actions in the service processes are preferred in clean production.

Clean production is closely related to the concept of sustainability, as well as to the development of environmentally friendly products, processes, systems, and services that are compatible with natural processes (Ljungberg, 2007). The United Nations Environment Programme (UNEP) notes that the clean production approach plays an important role in fulfilling the responsibilities of countries party to international agreements (Manzini et al., 2001).

While many businesses have existing internal audit systems, the size of the audit is important when the audit subject is the environment. As to compliance with international standards (such as BS7750-EMAS and ISO 14001), relevant authorities and government agencies view the environment as an international public property. As such, relevant audits include facility, environmental impact, compliance, minimization, property transfer, waste, and life cycle (Dittenhofer, 1995).

CLEAN PRODUCTION AND GOVERNMENTS

Survivors of pollution after the end of economic activity are only marginally endogenous. Preventing externalities resulting from the production process often yields benefits, but such a change might be costly. At this point the objective is to keep external costs low (Rousse, 2008).

Failing states and the over-developed public economy, which result from today's accepted social welfare state, are optimizing the distribution of resources and thus preventing society from achieving maximum refinement. In response, the idea of the optimal (limited liability) state, designed to address these problems, has been proposed. The optimal state is a limited and responsible state, affording rights and freedoms to a person who does not negatively intervene (Beck, 2000). Under this approach, the duties and functions of the state are defined in the relevant state constitutions. The state takes the role of referee in the market economy, and the players are able to play the game under the rules of the game. States intervene in a limited fashion, only when necessary (Rapaczynski, 1996).

When the subject is the environment and clean production, in addition to limited state intervention, there is also a need for guiding duties and policies. Otherwise, the costs for reducing the negative externalities from pollution mentioned above will be permanently postponed by the market. Without government intervention, environmental regulations imposed on manufacturers and the market are fairly limited. This leads to environmental pollution as there are no government audits. Clean production goals over the period 2007-2013 were:

1. To prevent air, water, soil, and marine pollution by taking into consideration the receiving capacity of the environment in the process of decreasing the effects of industrial facilities, the pollution load, and the establishment of new facilities.
2. To provide effective waste management.

In order to increase environmental taxation in developing countries, it is necessary to carry out the process of interaction between Economic sectors in a highly planned and healthy manner, and to implement a structure of necessary public policies within the strategic construction timing (Yeldan et al., 2013). The share of environmental taxes in other tax revenues increases only when the environmental awareness and economic well-being of that country increases. This points to the importance of implementing constructive guiding activities.

CLEAN PRODUCTION AND GREEN GROWTH RELATION

The current perception of sustainable development has evolved the concept of growth to the concept of green growth, especially in developed countries. The concepts of intelligent growth, inclusive growth, green growth, and sustainable growth, expressed within the context of the European Union 2020 Strategies, represent a new approach to the growth dimension of developed countries (Sarkar, 2013). These growth priorities are 21st century, and a new vision for the European social market economy is proposed (Sarkar, 2013).

Green growth is exhibited in existing sustainable development initiatives in many countries. The aim is to identify cleaner growth sources, including the use of new green industries and business and technology development opportunities, while concurrently managing structural changes related to a green economy transition.

DISCUSSIONS AND CONCLUSION

There has been consensus among participants in the public debate on the necessity of state subsidies and sanctioning power due to cleaner transit costs in order to prevent environmental negative market externalities when the issue is the environment and externalities. Although consensus may change if the interview content is widened to include more sections, the general opinion is that the transition to clean production can only be possible with state support and sanctions.

Subsidies and other support policies with a focus on economic development (including grants and funds) increase the demand for green production preferences in the market, including by:

- Promoting research, development, and innovation to raise environmental investment rates.
- Prioritizing environmental entrepreneurship; supporting research and innovation-based project markets in a centralized database to support the dissemination of successful implementations across the country in the framework of intellectual property and copyright; collaborating with universities; and establishing a clean production transition phase in public policy, as the one in Canada supported by commissions (to develop analyses, reports, and strategies) to be carried out by private sector / university collaborations, such as "Green Budgeting Commission" and "Green Economy Transition Commission."
- Implementing good governance in the context of openness, accountability, management ethics, rules and restrictions, on-the-ground management, digital revolution compliance, functioning of regulatory and market mechanisms, rule of law, auditing, effective civil society, and participation.
- Ensuring that subsidies are consistently applied to develop the economic segment where environmental taxation is to be widespread and contributing to reducing negative externalities on the environment.
- Accelerating the transition of the public share in production to the clean production system.
- Initiating "Clean Production Specialist" training programs at vocational schools.
- Expanding the policies necessary for the Waste Stock Exchange to be activated.
- Developing legislation on pollution prevention approaches in production, with an emphasis on eliminating legal gaps.

For companies operating in complementary industrial sectors where the organized industrial zone approach will be entirely a raw material-waste-to-industry chain, and where the raw material input and waste rates can be monitored, that software be developed to prioritize this conversion approach. In this context, the requirements for clean production are particularly important in terms of policies that will be produced, especially in R & D and technology:

- Innovative clean (sustainable) production practices in priority sectors and detailed product research;
- Investigating the applicability of specific clean (sustainable) production tools;
- Research on increasing environmental performance in various sectors;
- Research on environmentally sensitive equivalents of polluting industry chemicals;
- Investigating waste products, energy, and bio-products from waste;
- Investigating industrial processes, product quality, and cost-effectiveness of clean (sustainable) production methods to be used in all sectors, including priority sectors;
- Investigating how existing waste management facilities can be used for cleaner (sustainable) production;
- Nano-technology, biotechnology, information technology, and more research on the use of cleaner (sustainable) production-based use of emerging science branches, in particular for our country; and
- To evaluate the transfer of clean technologies on a sectoral basis and to cover university-private sector cooperation (Tütüncü & Şahin, 2012). This includes the effective use of environmental indicators, transition to environmental accounting systems, cleaner production, pollution prevention, industrial ecology, livelihood assessment, eco-efficiency, environmental design, and green logistics and green marketing, to take an innovative approach to the regional development objective dimensions, and an innovative approach to environmental and industrial environmental management processes should be prioritized in the context of state support (Lozano, 2012).

Given the threatening position of sustainable growth on the balance of global ecology and economic growth ambitions, a focus on increasing societal spending nowadays to address the negative externalities of economic activity inevitably contributes to the formation of large environmental crises behind economic crises. Systematic productivity and quality-oriented policies that closely monitor European and world applications and that reflect on their own financial transformation strategies are necessary. Not only must these policies be in accordance with international treaties, but also with environmental financial practices, and they must strengthen political and economic structures.

ORGANIZATION OF THE BOOK

The book is organized into nine chapters. A brief description of each follows:

Chapter 1 discusses the United Nations World Water Day, global water problems, and water pollution prevention, and a researcher-oriented platform is established.

Chapter 2 presents the effects of fossil fuel emissions on air pollution, especially in metropolises, and solutions proposed to mitigate such damage are discussed. The causes of diseases in human respiratory tracts is also discussed.

Chapter 3 discusses chemical precipitation and scrubber systems for removing emission gases such as SOx, NOx, and COx emitted from acid and fertilizer production facilities, and alternatives are shown.

Chapter 4 discusses the excellent adsorbent properties of SBA-15 mesoporous silica material for CO_2 emission reduction. The behavior of a modified SBA-15 adsorbent medium is described.

Chapter 5 presents a local municipality study on chemical characterization for waste disposal, as well as research on solid waste management systems.

Chapter 6 proposes a new method called exposure analysis for measuring pollution levels, and health effects are discussed when the pollution concentration is estimated. In addition, exposure analysis was performed for open and closed environments.

Chapter 7 discusses the increase in economic purchasing power of people in relation to the increase in agriculture and industrial production. With an increase in industrial production, an increase in water and air pollution has become seemingly inevitable. Measures taken to prevent increases in water and air pollution are also discussed.

Chapter 8 evaluates the performance of renewable energy sources in preventing water and air pollution. During this evaluation, relevant parameters were investigated and the effects on performance are discussed.

Chapter 9 discusses the impacts of global warming on climate change and the effects on human health and sea level. While discussing these possible effects, the applicability of the methods developed to prevent them is questioned.

Ahmet Ozan Gezerman
Yildiz Technical University, Turkey

REFERENCES

Beck, U. (2000). *The brave new world of work.* Cambridge, UK: Polity Press.

Dittenhofer, M. (1995). Environmental accounting and auditing. *Managerial Auditing Journal, 10*(8), 40–51. doi:10.1108/02686909510093615

Ljungberg, L. Y. (2007). Materials selection and design for development of sustainable products. *Materials & Design, 28*(2), 466–479. doi:10.1016/j.matdes.2005.09.006

Lozano, R. (2012). Towards better embedding sustainability into companies' systems: An analysis of voluntary corporate initiatives. *Journal of Cleaner Production, 25*, 14–26. doi:10.1016/j.jclepro.2011.11.060

Manzini, E., Vezzoli, C., & Clark, G. (2001). Product-service systems: Using an existing concept as a new approach to sustainability. *Journal of Desert Research, 1*(2), 27–40.

Rapaczynski, A. (1996). The roles of the state and the market in establishing property rights. *The Journal of Economic Perspectives, 10*(2), 87–103. doi:10.1257/jep.10.2.87

Rousse, O. (2008). Environmental and economic benefits resulting from citizens' participation in CO_2 emissions trading: An efficient alternative solution to the voluntary compensation of CO_2 emissions. *Energy Policy, 36*(1), 388–397. doi:10.1016/j.enpol.2007.09.019

Sarkar, A. N. (2013). Promoting eco-innovations to leverage sustainable development of eco-industry and green growth. *European Journal of Sustainable Development, 2*(1), 171.

Tütüncü, S. İ., & Şahin, N. (2012). Üretimde Kirlilik Önleme Yaklaşımı ve Devletin Konumu. *Ekonomi Bilimleri Dergisi, 4*(2).

Yeldan, E., Taşci, K., Voyvoda, E., & Özsan, M. E. (2013). Turkey on her way out of middle-income growth trap. Report for the Turkish Enterprise and Business Confederation.

Chapter 1
A Framework for Assessment of Existing Solid Waste Management Practices and Characterization of Municipal Solid Waste in Muzzafarnagar City, India

Ankur Choudhary
Jaypee University of Information Technology, India

Rajiv Ganguly
Jaypee University of Information Technology, India

Ashok Kumar Gupta
Jaypee University of Information Technology, India

ABSTRACT

This chapter reports the details of the existing system of MSW management and characterization of Muzaffarnagar City located in Western Uttar Pradesh (UP) state in India. The overall waste generated in the city is about 120-125 tons per day (TPD) with a per capita generation rate of 0.415 kg/person/day with a collection efficiency of 70-80%. Physico-chemical and geotechnical properties of the MSW were carried out to determine its overall characteristics. The characterization results showed about 46% of the waste generated in the city is organic nature (from HIG and MIG) and 52% for (LIG) with chemical characterization showing that the elemental carbon was in the highest proportion. Further, the chapter also recommends suitable remedial measures for proper management of the existing MSW management system and suitable treatment alternatives.

DOI: 10.4018/978-1-5225-3379-5.ch001

INTRODUCTION

Rapid rise in India's economy due to increased urbanisation and globalization and thereby changing lifestyle coupled with increasing population (births, migration etc.) has led to production of eight fold increase in MSW generation in comparison to 1947 (Sharholy et al, 2008; Talyan et al, 2008) leading to severe degradation of existing environmental conditions (Hazra et al, 2009). With the passage of time it is expected that Indian population will increase further with the annual rate of growth of urban population in India being predicted at 5% ("Census of India", 2011). Further, rapid influx of population from rural areas in urban cities lead to unwanted population growth leading to development of unplanned rural areas in the outskirts of the Indian cities leading to additional generation of MSW which are often unaccounted. In general, waste is defined as a material of unuseful nature and of no economic value to its owner, the owner being the generator of the waste (Mor et al, 2006b). Effective management of MSW is a global problem with developing countries facing the biggest obstacles (Ramachandra et al, 2003; Tchobaanpoglous et al, 1993). It has been inferred that increasing population is directly related to rapid increase in MSW generation in developing countries with India being no exception and might be reaching critical levels particularly due to unavailability of barren land for disposal (Indris et al, 2004; Talyan et al, 2008).The per capita waste generation of India rate varies between 0.15 kg in rural areas to 0.45 kg in urban areas (Akhtar, 2014; Katiyar et al, 2013) and it has been observed that increased economic success leads to increased per capita consumption and generation of MSW (Taylan et al, 2007; Tricys, 2002). The average per capita waste generation in India is 370 g/day as compared to 2200 g/day in Denmark, 2000 g/day in US and 700 g/day in China (Liu et al, 2011).

In an Indian context, the MSW can be classified primarily as the solid waste generated from domestic, commercial, institutional sources and to a lesser extent biomedical and industrial (toxic waste). The most prevalent method for disposal of such wastes in India are open dumping (90%) which over the period of time leads to choking of sewer pipes, breeding grounds for vectors and other serious heath hazardous and environmental problems. Such status of MSW disposal is regularly observed for Tier-II and Tier-III cities. In Tier-I and metropolitan cities some of which have implemented management strategies for tackling such huge voluminous quantities the procedures implemented have been outdated and inefficient due to lack of characterization studies which would have enabled these municipalities in implementing an effective treatment and disposal system. For example, it is estimated that about 1.3 million cubic metre of biogas per day or 72 MW of electricity from biogas can be produced for agricultural purposes along with about 5.4 million metric tonnes of compost annually ("Central Pollution Control Board", 2015). However, due to improper MSW management practices including no segregation of wastes being followed the end products of several waste to energy (WTE) systems implemented are poor in nature with no economic value (Rana et al, 2015).

In the above context, it is an essential duty of concerned urban local body (ULB) for effective management of the MSW generated within their city limits ("The Gazette of India" 2000). However, implementation of an effective MSW management system requires suitable data including quantity, quality and characterization results which are further dependent on geographical locations, food habits and other socio-economic conditions and also adequate budgetary provisions (Goswami et al, 2007). A detailed characterization study can help in determining the drawbacks in the existing treatment processes; identify the most appropriate treatment technology for effective management of the MSW

The main objectives of the study were to assess the existing MSW management and determine the characteristics of the waste generated in Muzaffarnagar city to determine deficiencies in the existing management system and propose remedial methods for improving the existing system.

MUZAFFARNAGAR CITY AND EXISTING MSW MANAGEMENT

Muzaffarnagar Municipality has a population of 392,451 and the urban agglomeration has a population of 4, 95,543 as per census 2011 with a growth rate of about 17% in a decade (2001-2010) covering an area of 150 Km^2 ("Census of India" 2011). The population of the city is increasing at a rate of 1.5% per annum. Presently, the city is divided into 41wards. The overall municipal solid waste generation in the city is about 120-125 Tonnes/day (TPD) [Personal communication with an employee of A2Z Infrastructure Limited, 2017] with an average generation rate of 0.32 kg/(capita-day). The per capita generation can be further subdivided for different socio-economic groups with average per capita generation of 0.14, 0.17 and 0.65 kg/day respectively for Low Income Group (LIG), Middle Income Group (MIG) and High Income Group (HIG) respectively.

The city has ample number of institutes, hospitals, shops, hotels and restaurants and approximately 64000 households [Personal communication with an employee of A2Z Infrastructure Limited, 2017]. Further, biomedical waste generated from hospitals and clinics (particularly syringes) do not undergo separate processing. These biomedical wastes are mixed with the MSW and are dumped at the non-engineered landfill site. Moreover, manufacturing wastes generated from different industries including sugar mills, paper mills and Steel Rolling mills don't under go to any special treatment and are disposed with the MSW making the overall nature of the waste highly hazardous.

Collection efficiency of domestic MSW generated is about 70-80% of the city is covered by door-to-door collection method using handcarts of volume 2-5 m^3. As mentioned earlier, the city is divided into 41 wards and if certain wards are not covered by door-to-door collection these wards have a designated collection point for waste. Insufficient bin capacity, use of open bins and location of these bins in inaccessible stations in some wards lead to overflowing of such bins almost through the entire year. Further, some wards have no designated bins leading to open dumping of generated MSW in these wards. No source segregation is followed and hence no separate bins exist for organic and inorganic wastes. Further, it has been observed that waste is open dumped in middle of road particularly in the vegetable market areas and even though collection is done on a daily basis if there is a lapse in collection of one day could lead to epidemic like conditions.

Loading of MSW is done manually and there exists only one garage in the city from where the transportation vehicles are operated and maintained with the help of private contractors. Transportation of the collected MSW is done using a total of 13 collection vehicles (6 dumpers, 3 J.C. Bamford Excavators limited, popularly designated as JCB, 3 tractors and 1 Tata Ace). Assuming the density of waste to be around 300 kg/m^3, the collection capacity per capita was determined to be only 0.001 m^3 which is extremely low and almost negligible highlighting the need for procurement of additional transportation vehicles. Further, lack and too much dependency of labourers, unplanned routing, non-covering of waste on top during transportation process highly reduces the efficiency of the transportation system. There are no intermediate transfer stations and no equipment or machinery is present at the open landfill site for segregation of wastes due to which mountains of waste exist in and around the landfill site and since last few years the load on the dumpsite has increased significantly. This leads to production of highly obnoxious odours, breeding ground of infectious vectors and propagate unhygienic environment in the vicinity of the dumpsite. Since, it is a non-engineered landfill contamination of groundwater and soil due to leachate percolation could lead to severe environmental hazard.

Though Ministry of Environment and Forests (MoEF) has specified detailed guidelines for management of MSW but majority of the cities in India generally follow only comprises only four activities (waste generation, collection, transportation and disposal) for management of MSW which is inadequate (Sharholy et al, 2008). The same was observed for our study location. Figure 1 shows the study location and Figure 2 summarizes the existing MSW management practice at the study location. Though there is an integrated SWM facility in the city, very poor condition of plant including machinery, equipment, scarcity of technical manpower, lack of interest of local bodies, lack of detailed characterization results, limited funds and inadequate management procedures not being thereby reducing the efficiency of the process. In

Figure 1. Location map of Muzaffarnagar, Uttar Pradesh

Figure 2. Schematic diagram of Solid Waste Management Practice in Muzaffarnagar city

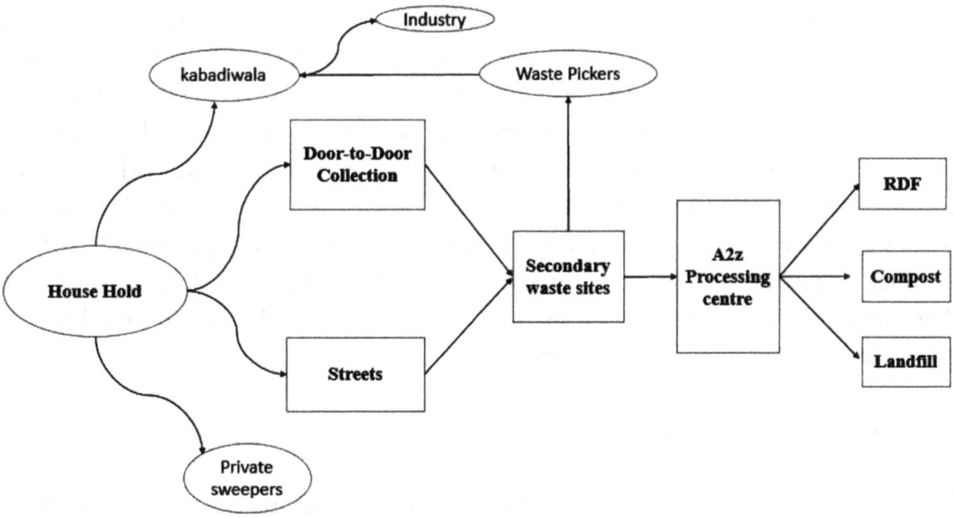

practice, since waste generation rate is very high and capacity of machines installed to processes the waste is very low, the waste stream coming to the plant is being underutilized leading to waste being accumulated.

MATERIAL AND METHODS

Sample Collection

Sampling was done as per the guidelines laid out in ASTM D5231-92 (ASTM, 2004a, ASTM, 2004b; ASTM 2004c; ASTM 2006a; ASTM, 2008. To summarize, sampling was carried out consecutively for a week at the dump site, with each day 150 kg sample collected while trucks used to unload. From the total waste collected over the week, about (n =7 (denoting sampling days); collection =1050 kg) about 100 kg of the final homogenised mixed waste was processed. Further, the physical characterization was carried out for LIG, MIG and HIG, to determine the characteristics of the MSW generated from the three different socio-economic groups.

Physical Characterization of MSW

Physical characterization is one of the most important components for designing an effective MSW management system. The samples (collected using the procedure described above) was segregated manually into food waste, garden trimming, paper, plastic, rubber, glass, metals, drain silt and construction debris and weighted to express as percentage of total weight. The fundamental aim of following this process was to identify the organic and the inorganic components. One of the major sub-classification of inorganic waste type is inert waste the major characteristics of which are non-reactive and non-biodegradable. For our present study, the inert fraction was primarily due to stones, sand and gravel. Such high quantities of inert wastes are generally accumulated due to road sweeping of construction dust which is collected and dumped at the open landfill site. Moisture content of the waste was determined immediately after collection of the samples and then the samples were processed for chemical characterization.

Chemical Characterization of MSW

Chemical characterization was carried out using both proximate and ultimate analysis to determine the elemental composition of the MSW and also its ash content. Collected samples were processed as per the procedure laid out in BIS-IS: 9234 and ASTM-D5231-92 (ASTM, 2004a; ASTM, 2004c; ASTM, 2006a; ASTM,

2008). While the proximate analysis method was adopted to determine ash content, fixed carbon and volatile matter the ultimate analysis was used to determine the proportions of carbon, hydrogen, nitrogen, sulphur and oxygen content in dry sample. The calorific value of the MSW waste sample was determined by bomb calorimeter (Model 6200 Spectronics).

Geotechnical Characterization of MSW

Different sieves with diameter 200, 150, 100, 50 and 20 mm was used to perform the gradation of MSW. Generally, geotechnical properties (moisture content, specific gravity, unit weight, field capacity, coefficient of permeability, and degree of saturation) helps in design and maintenance of the landfill. Constant head method has been used to evaluate the coefficient of permeability at 25^0C in the laboratory.

RESULTS AND DISCUSSION

The characteristics of a MSW generated at a location are significantly influenced by geographical demographics, lifestyles, cultural traditions, economic prosperity, literacy rates, dietary habits and climatic conditions (Jin et al, 2006). Accurate characterization of MSW generated is the foremost step in designing an effective management system (Lin et al, 2011; Rana et al, 2014).

Physical Characterization

Physical characterization is helpful in determining the organic and inorganic fraction present in the waste stream and forms the basis for determining the treatment alternative. The results of physical characterization of MSW from Muzaffarnagar have been summarized in Table 1.

Density of MSW helps in deciding the appropriate handling processes including collection and transportation of the MSW generated. A higher inert fraction as observed from Table 1 significantly increases the bulk density and was found to 475 kg/m^3. Further, it is observed that the organic fraction accounts for the highest proportion from the entire socio-economic group (about 46% from HIG and MIG) and about 52% from LIG followed by inert constituents (27%, 34% and 32% for HIG, LIG and MIG respectively).

As observed from Table 1, the organic fraction comprises mainly of food wastes and garden trimmings for the different socio-economical conditions. Further, the maximum organic fraction is generated from the LIG community and is similar to other reported studies. This is primarily because with decrease in economic status

Table 1. Physical Characterization of Municipal Solid Waste in Muzaffarnagar city

Waste Type	HIG (%)	MIG (%)	LIG (%)	Mean± SD (%)
Food Waste	14.82	17.26	24.72	18.93±5.16
Garden Trimming	31.45	29.12	27.56	29.38±1.96
Paper	7.88	5.12	5.23	6.08±1.56
Plastic and Rubber	17.23	13.76	9.09	13.36±4.08
Glass	0.7	0.45	0.32	0.49±0.19
Metals	0.45	0.35	0.28	0.36±0.09
Drain Silt	21.24	25.34	26.43	24.34±2.74
Construction Debris	6.23	8.6	6.37	7.07±1.33

of the location there is an increase in the organic content of the waste generated (Goel, 2008; Sharholy, 2008). Further, it has been reported that MSW generated in developing countries have high proportion of organic waste (40-70%) making them unsuitable for disposal by incineration (Khajuria et al, 2010).

It has been further reported that proportion of paper increases with increase in population and generally varies between 1 to 6 (Garg et al, 2012). For our study location, the fraction of paper constituent was determined to be 5.3%, and lies within the range as reported by literature. In addition, the combustible fractions of the waste including paper, pouches, cardboards, polyester fibres, rubber, leather, egg tray, jute bags etc. was determined to be about 19%. Such other studies carried out in Indian context, reported similar values for Varanasi with combustible fraction being 14% (Dasgupta et al, 2013). High proportion of plastic and rubber was also observed as plastics are used on daily basis by the local residents. In particular, plastic, rubber and paper waste were significantly higher in HIG in comparison to LIG and MIG areas, similar to other reported literature. Glass and metal fraction was higher in HIG as compared to the other socio-economic groups.

Inert fraction was determined to be approximately 35%, which mainly include sand, silt, ash, and was on the higher side. This is primarily due to the street sweeping of construction wastes which end up being finally disposed off at the landfill site. The inert fraction determine for Muzaffarnagar was similar to other reported studies in Indian cities, for example Pune reported about 26% of inert wastes (Mane et al, 2012); Jalandhar reported about 21-33% of inert (Sethi et al, 2013) and the tricity regions of Chandigarh, Mohali and Panchkula as about 25-30% (Rana et al, 2017). Further, increased inert fraction decreases the calorific value and increases the density of the waste (Sethi et al, 2013). Drain silt fraction was found to be maximum in LIG areas because unpaved roads are more in the LIG locations in comparison to HIG areas which conditions are roads which have more paved roads reducing the percentage of drain silt.

Recyclable components include glass, metal, plastic, cartons packs, synthetic fibres (nylon ropes), were determined to be approximately 1%, which is very less, primarily due to rag pickers and waste collectors who cover all of residential areas, markets institutes, industrial areas collecting all recyclables and then selling it to the *kabaddis*. They directly sell these to the industries to be recycled.

Chemical Characterization

Physical Characterization of the MSW provides information of the components of the waste stream but is not sufficient enough for deciding the overall processing technique of the waste, hence chemical characterization is performed in order to determine crustal element present in the MSW. Ultimate and Proximate analysis were carried out to determine the chemical characterization and the results have been summarized in Table 2.

Moisture content in Asian countries vary from 17-65% (Kolekar et al, 2016; Kumar et al, 2009) and the moisture content of our study location was determined to be about 22.6% lying within the prescribed limits. High value of moisture content indicates higher fraction of biodegradables.

Table 2. Chemical Characterization of Municipal Solid Waste in Muzaffarnagar city

Parameter	Mean± SD
Moisture Content	22.60±1.16
Volatile Matter	22.30± 2.63
Ash Content	44.90± 2.30
Fixed Carbon	10.10 ±1.57
Calorific Value	4236.15 ±119.78
Carbon	31.21±2.35
Hydrogen	3.87±0.52
Nitrogen	1.34±0.29
Oxygen	20.91±0.63
Mineral Content	42.68±2.29
C/N	23.75 ±4.78

*All values in percentage except calorific value (kcal/kg) and C/N

Ash Content was determined to be 44.9% primarily due to the high fraction of inert present in the waste stream. Such high percentage of ash content has also been reported for earlier studies in Jalandhar (42%) (Rawat et al, 2013) was reported.

The C/N ratio for Muzaffarnagar was determined to be 23.75 which were well within the C/N ratio reported for Asian countries which vary from 17-50 (Kolekar et al, 2016; Rawat et al, 2013; Sharholy et al, 2008). Elemental composition shows a high fraction of C (31.21%) followed by O (20.9%), H (3.87%), N (1.34%). Analysis of elemental components are particularly useful in determination of stoichiometric reactions and thereby estimation of gaseous by products of the waste (Khajuria et al, 2010).

Determination of Calorific Value provides information regarding energy recovery from the waste. Due to a high proportion of combustibles (19%), the calorific value of the waste generated from Muzaffarnagar area has been determined to be 4236 kcal/kg. In practice, calorific value generally varies between 800 to 1000 kcal/kg (Saha et al, 2010). However, studies conducted at Indian cities show higher reported calorific values.

Geotechnical Characterization

Particle size distribution of the different components has been summarized in Table 3.

It is observed from Table 3 that paper and cardboard has the maximum size range whereas ash and dust have particle size less than 20 mm and the food component varies between 20-100 mm. Particle size distribution of different waste stream characteristics are important as decomposition rate is directly proportional to the size distribution. Further, lesser the size of the particle lesser will be the pore size between two adjacent particles which limits the passage of oxygen, thereby facilitating the process of anaerobic degradation. The optimum particle size should vary between 12 to 50 mm for design of a successful composting system (Thakur et al, 2015).

Table 3. Particle Size distribution of Municipal Solid Waste in Muzaffarnagar city

Component	Size Range (mm)
Food	20-100
Paper and Cardboard	above 200
Plastic	50-150
Glass	50-200
Rubber	below 20
Ash and Dust	below 20
Rags	100-20

Geotechnical properties of the waste affect waste degradation, leachate distribution, and the overall stability of landfills (Yu et al, 2011). The geotechnical characterization of the MSW has been summarized in Table 4.

The moisture content was determined to be 0.19, which is very less in comparison to other reported literature (Sethi et al, 2013; Rana et al, 2017). This was primarily due to the climatic conditions existing at the study location which is exceedingly hot and had experienced very less rainfall. The unit weight was determined to be 11.34 kN/m^3. This parameter is primarily significant in determining the strength of the landfill as it depends upon the effective over burden pressure and effective overburden pressure is directly related to the unit weight. Further, it has been reported that unit weight also effects the capacity of the landfill (Yu et al, 2011). The degree of saturation of the waste was determined to be 76% and the coefficient of permeability was determined to be 3.0 x10^{-3} cm/s.

RECOMMENDATIONS FOR IMPROVEMENT OF EXISTING MUNICIPAL SOLID WASTE MANAGEMENT PRACTICES BASED ON CHARACTERIZATION ANALYSIS

The characterization study carried out for Muzaffarnagar reveals that an integrated MSW management system is of immediate need. Though there exists waste processing system, it is severely lacking and needs a severe upgradation or a complete overhaul. The following remedial measures are suggested for better management of the MSW generated in Muzaffarnagar.

Table 4. Geotechnical Properties of Municipal Solid Waste in Muzaffarnagar city

Parameter	Mean± SD (%)
Moisture Content (v/v)	0.19±0.03
Degree of Saturation (%)	76.1±9.6
Specific gravity	2.15±0.11
Wet Unit Weight (kN/ m^3)	11.34±1.82
Coefficient of Permeability(cm/s)	3 x 10^{-3}

Source Segregation

Physical characterization of the MSW carried out has revealed that about 48% of waste stream are biodegradables and about 32% of inert materials. In this context, source segregation of the waste should be carried out to effectively reduce the burden on the landfill site and for implementation of adequate treatment procedures. Further, unsegregated wastes significantly reduce the calorific value of the waste thereby diminishing the potential energy content. The process of source segregation can be easily implemented with provisions of two bins (or colour coded containers) one for collection food waste and other organic materials and the second bin for other waste types. Further, education and importance of source segregation should be imparted to the local residents to apprise them of the potential benefits of the process. It had also been observed from our study that the combustible fraction is about 19% and it would be highly beneficial if source segregation of combustible methods were performed to yield better quality of Refuse Derived Fuel (RDF). In particular, it should be ensured that inert waste should not be mixed with domestic generated solid waste. The density of the MSW was determined to be 475 kg/m^3 and an effective strategy would be implement adequate size and number of bins depending upon the population, waste generation rates and density of wastes at the different wards of the city. The number of transportation vehicles should be increased and proper maintenance along with planned route layouts should be followed to improve the efficiency of the disposal system.

Recycling and Recovery

Our study show that about less than 1% of recyclables enter the waste stream with informal recycling process being most predominant carried out by rag pickers and waste collectors which are passed onto the industries. Though to a certain extent, such informal recycling practices reduce the burden on the landfills, the absence of formal recycling units significantly lower the economic potential of the recyclables with no benefits being passed onto the government. In this context a formal recycling unit should be set up with proper management to derive all potential benefits of the recycling process.

Installation of Biomethanation Plant

Since, the municipal solid waste consists of a high fraction of biodegradables in the waste stream it will be highly appropriate to install a biomethanation plant. Further, organic wastes from other sources should also be utilized (like kitchen and restaurant wastes) for proper functioning of the biomethanation plant.

Composting

The climatic conditions of the study location are highly suitable for composting process. Composting process is one of the finest methods for disposal of organic waste as it stabilizes the waste and the end product of the composting process can be used as a natural fertilizer. It is an environmental friendly aerobic process carried out through microbial action. The efficiency of the composting process depends upon the efficiency of segregation of organic wastes from other types. Further, composting process can be carried out at household levels and as such proper training and education should be provided to the residents to start their own household composting systems. At the governmental levels, community composting pits should be set up to utilize the voluminous quantity of organic waste generated.

Installation of RDF Plant

It is observed from our characterization study that a RDF could be utilized effectively as a part of an integrated MSW management system. In practice, a working RDF plant already exists in Chandigarh city and can be similarly implemented for Muzaffarnagar. For the existing plant in Chandigarh it has the capacity to treat about 500 tonnes per day (TPD) with the RDF produced having a calorific value of 3100 kcal/kg and moisture content less than 15%. Implementation of RDF unit will significantly reduce the volume of waste thereby increasing the lifespan of the landfill.

Engineered Landfill and Other Recommendations

Municipal Solid waste is openly dumped at the landfill site leading to possible contamination of soil and groundwater. In this context, it is proposed a proper engineered landfill site with leachate collection and extraction system should be installed to avoid such contamination. Skilled manpower should be employed and periodical trainings should be provided to such personals for operation of such systems. Further, the personals should be educated on occupational health hazard and medical checkup of ragpickers and waste collectors to be done.

CONCLUSION

The book chapter reports the existing solid waste management practices in Muzaffarnagar city along with characterization analysis studies which are then utilized in recommendations as important strategies for the municipal solid waste management. With an increased population and excessive consumption rates the city is nearing a critical point with no implementation of an integrated MSW management facility. Existing status for management of MSW is deficient in almost all aspects with less collection efficiency due to inappropriate bin size, inaccessible locations of bins, inadequate number of transportation vehicles per and non-functioning of the integrated MSW management system. Physical Characterization of the waste stream generated from the Muzaffarnagar denotes high fraction of organics followed by inert. Waste streams from LIG were determined to have the highest organic fraction. Chemical characterization showed a high proportion of ash content primarily due to presence of inert. In this context, source segregation should be carried out to obtain maximum benefits of using WTE technology with biomethanation and composting for organic fraction and RDF for the inorganic fraction of the waste. Further, a formal recycling unit should be setup to to derive the economical benefits of the recycling system. In particular, local administration will need full support of the local public to implement any potential schemes for integrated MSW management system. This can be done by educating the general public on the drawbacks of the existing waste management practices. There is a great scope for improvement in the existing practices of MSW management in the city, however significant efforts are required by the Urban Local body of Muzaffarnagar and support of the public. Waste reduction, waste segregation at source, reuse, recycling, and WTE schemes like composting, biomethanation and RDF could provide long-term solutions to reduce the burden of waste disposal on the landfill site. Finally, the open dumpsite should be converted to an engineering landfill site with immediate effect.

REFERENCES

Akhtar, M. N. (2014). Prospective Assessment for Long-Term Impact of Excessive Solid Waste Generation on the Environment. *International Journal of Advancement in Earth and Environmental Sciences*, *2*(2), 39–45.

ASTM. (2004a). *Standard test method for residual moisture in arefuse-derived fuel analysis sample. E790*. West Conshohocken, PA: ASTM International.

ASTM. (2004b). *Standard test method for ash in the analysis sample ofrefuse-derived fuel. E830*. West Conshohocken, PA: ASTM International.

ASTM. (2004c). *Standard test method for volatile matter in the analysis sample of refuse-derived fuel. E897*. West Conshohocken, PA: ASTM International.

ASTM. (2006a). *Standard test methods for specific gravity of soil solids by water pycnometer. D854*. West Conshohocken, PA: ASTM International.

ASTM. (2008). *Standard test method for determination of the compositionof unprocessed municipal solid waste. D5231-92*. West Conshohocken, PA: ASTM International.

Census of India. (2011). New Delhi, India: Ministry of Home Affairs, Government of India.

Central Pollution Control Board (CPCB). (2015). *Status of solid waste generation, collection, treatment and disposal in metro cities*. Delhi: Author.

Dasgupta, B., Yadav, V. L., & Mondal, M. K. (2013). Seasonal characterization and present status of municipal solid waste (MSW) management in Varanasi, India. *Advances in Environmental Research*, *2*(1), 51–60. doi:10.12989/aer.2013.2.1.051

Garg, V., & Rani, J. (2012). Perspectives of Municipal Solid Management In India: A Case Study Of Chandigarh. *International Journal of Applied Engineering Research*.

Goel, S. (2008). Municipal Solid Waste Management (MSWM) in India - A critical Review. *Journal of Environmental Science & Engineering*, *50*(4), 319–328. PMID:19697768

Goswami, U., & Sharma, H. P. (2007). Study of groundwater contamination due to municipal solid waste dumping in Guwahati city. *Pollution Research*, *26*(2), 211–214.

Hazra, T., & Goel, S. (2009). Solid waste management in Kolkata, India: Practices and challenges. *Waste Management (New York, N.Y.)*, *29*(1), 470–478. doi:10.1016/j.wasman.2008.01.023 PMID:18434129

Indris, A., Inane, B., & Hassan, M. N. (2004). Overview of waste disposal and landfills/dumps in Asian countries. *Journal of Material Cycles and Waste Management*, *6*, 104–110.

Jin, J., Wang, Z., & Ran, S. (2006). Solid waste management in Macao: Practices and challenges. *Waste Management (New York, N.Y.)*, *26*(9), 1045–1051. doi:10.1016/j.wasman.2005.08.006 PMID:16253497

Katiyar, R. B., Suresh, S., & Sharma, A. K. (2013). Characterization of municipal solid waste generated by city of Bhopal, India, ICGSEE-2013. *Proceedings of International conference on global scenario in Environment and Energy*, *5*(2), 623-628.

Khajuria, K., Yamamoto, Y., & Morioka, T. (2010). Estimation of municipal solid waste generation and landfill area in Asian developing countries. *Journal of Environmental Biology*, *31*, 649–654. PMID:21387916

Kolekar, K. A., Hazra, T., & Chakrabarty, S. N. (2016). A Review on Prediction of Municipal Solid Waste Generation Models. *Procedia Environmental Sciences*, *35*, 238–244. doi:10.1016/j.proenv.2016.07.087

Kumar, S., Bhattacharyya, J. K., Vaidya, A. N., Chakrabarti, T., Devotta, S., & Akolkar, A. B. (2009). Assessment of the status of municipal solid waste management in metro cities, state capitals, class I cities, and class II towns in India: An insight. *Waste Management (New York, N.Y.)*, *29*(2), 883–895. doi:10.1016/j.wasman.2008.04.011 PMID:18595684

Liu, C., & Wu, X. W. (2011). Factors Influencing Municipal Solid Waste Generation in China: A multiple statistical analysis study. *Waste Management & Research*, *29*(4), 371–378. doi:10.1177/0734242X10380114 PMID:20699292

Mane, T. T., & Hemalata, H. (2012). Existing Situation of Solid Waste Management in Pune City, India. *Research Journal of Recent Sciences, 1*, 348-351.

Mor, S., Khaiwal, R., Dahiya, R. P., & Chandra, A. (2006b). Municipal Solid Waste Characterization and its Assessment for Potential Methane Generation: A Case Study. *Journal of Science of the Total Environment*, *371*(1-3), 1–10. doi:10.1016/j.scitotenv.2006.04.014 PMID:16822537

Ramachandra, T. V., & Varghese, S. K. (2003). Exploring possibilities of achieving sustainability in solid waste management. *Environmental Health*, *45*(4), 255–264. PMID:15527017

Rana, P. R., Yadav, D., Ayub, S., & Siddiqui, A. A. (2014). Status and challenges in solid waste management: A case study of Aligarh City. *Journal of Civil Engineering and Environmental Technology, 1*(4), 19–24.

Rana, R., Ganguly, R., & Gupta, A. K. (2015). An Assessment of Solid Waste Management System in Chandigarh City, India. *The Electronic Journal of Geotechnical Engineering, 20*(6), 1547–1572.

Rana, R., Ganguly, R., & Gupta, A. K. (2017). Physico-Chemical Characterization of Municipal Soild Waste from Tri-city region of Northern India – A case study. *Journal of Material Cycles and Waste Management*. doi:10.1007/s10163-017-0615-3

Rawat, M., Ramanathan, A. L., & Kuriakose, T. (2013). Characterization of municipal solid waste compost from selected from selected Indian cities- A case study for its sustainable utilization. *Environment and Progress, 4*(2), 163–171. doi:10.4236/jep.2013.42019

Saha, J. K., Panwar, N., & Singh, M. V. (2010). An assessment of municipal solid waste compost quality produced in different cities of India in the perspective of developing quality control indices. *Waste Management (New York, N.Y.), 30*(2), 192–201. doi:10.1016/j.wasman.2009.09.041 PMID:19857948

Sethi, S., Kothiyal, N. C., Nema, A. K., & Kaushik, M. K. (2013). Characterization of Municipal Solid Waste in Jalandhar City, Punjab, India. *Journal of Hazardous, Toxic and Radioactive Waste, 17*(2), 97–106. doi:10.1061/(ASCE)HZ.2153-5515.0000156

Sharholy, M., Ahmad, K., Mahmood, G., & Trivedi, R. C. (2008). Municipal solid waste management in Indian cities - A review. *Waste Management (New York, N.Y.), 28*(2), 459–467. doi:10.1016/j.wasman.2007.02.008 PMID:17433664

Talyan, V., Dahiya, R. P., & Sreekrishnan, T. R. (2008). State of municipal solid waste management in Delhi, the capital of India. *Waste Management (New York, N.Y.), 28*(7), 1276–1287. doi:10.1016/j.wasman.2007.05.017 PMID:17692510

Taylan, V., Dahiya, R. P., & Anand, S. (2007). Quantification of Methane emission from Solid Waste Disposal in Delhi. *Journal of Resources Conservation and Recycling, 3*(3), 240–259. doi:10.1016/j.resconrec.2006.06.002

Tchobaanoglous, H., Theisen, H., & Samuel, A. (1993). *Integrated Solid Waste Management*. New Delhi: McGraw-Hill, Inc.

Thakur, I., Ghosh, P., & Gupta, A. (2015). Combined chemical and toxicological evaluation of leachate from municipal solid waste landfill sites of Delhi, India. *Environmental Science and Pollution Research International*, *22*(12), 9148–9158. doi:10.1007/s11356-015-4077-7 PMID:25578612

The Gazette of India. (2000). *Municipal solid waste (management and handling) rules, Ministry of Environment and Forests (MoEF)*. Author.

Tricys, V. (2002). Research of leachate, surface and ground water pollution near Siauliai landfill. *Environmental Research, Engineering and Management*, *19*, 30–33.

Yu, L., Batlle, F., & Carrera, I. (2011). Variations of waste unit weight during mechanical and degradation processes at landfills. *Waste Management & Research*, *29*(12), 1303–1315. doi:10.1177/0734242X10394912 PMID:21339240

Chapter 2
Air Pollution in Asia and Its Effect on Human Health:
Air Pollution in Asia

Ahmet Ozan Gezerman
Yildiz Technical University, Turkey

Burcu Didem Çorbacıoğlu
Yildiz Technical University, Turkey

ABSTRACT

Although continuous efforts to monitor and mitigate air pollution are being made, it is still prevalent in most countries in the world. Major contributors include fossil fuel exhaust in metropolitan cities from industrial facilities and vehicular emissions. Use of renewable energy and natural gas have played a part in reducing air pollution; however, increasing populations, rampant urbanization, and industrialization, especially during winter months, have given rise to spikes in air pollution levels. Research shows that there is a close relationship between air pollution and mortality rates depending on respiration inadequacy. Studies show that contaminants increase respiratory afflictions in humans. Discontinuing use of fossil fuels, using appropriate burning techniques, and efficiency emission controls on vehicles have been proven to reduce air pollution levels.

DOI: 10.4018/978-1-5225-3379-5.ch002

INTRODUCTION

Effects of Air Pollution on Human Health

Air Pollution in the World

Air pollution is defined as the level of contaminants, such as SO_x, NO_x, and O_3, that have negative effects on health. This contamination interferes with natural atmospheric cycles and negatively affects public health. Over the last 30 years, pollution levels recorded in big cities have been high. Until 1980, it is observed that 1.3 billion people lived in highly polluted cities. Air pollutants originate from industrial facilities, vehicular exhaust, and fuel consumption. Carbon emissions are expected to increase by five-fold by the year 2030. Pollution has also increased in rural areas due to rapid urbanization and increases in energy consumption(Akimotoi, 2017).

Chief components of air pollution include aerosol, fume, black carbon, and dust. Most contaminants contain sulfuric acid, sulfate, and nitrate salts . These compounds significantly affect the respiratory tract in humans. SO_2 is a gaseous contaminant that is eliminated in the nose, while ozone and NO_2 being insoluble in water tend to respiratory system. CO is directly soluble in the blood (Bayram et al, 1998).

Several studies on the chemical effects of air pollution on health have been conducted in the last 20 years. They have revealed that air pollution increases respiratory infection risk in children (Bayram et al, 2006). The studies were initiated after pollution episodes in London in 1952, in Donora in USA in 1947, and Meuse valley of Belgium in 1934, resulted in significant casualties (Bayram et al, 1998). In response to these episodes, emission control approaches that limited the use of coal were enforced in London. Air pollution checks, emission criteria and limitations have been also enforced by the World Health Organization (WHO). Concentrations of $SO_2$and other selected gas contaminants as air quality criteria were selected by WHO, and have shown to be rather high in summer and winter months. Air pollution has recently been on the rise in Southeast Asian countries(Bayram et al, 2002).

Air Pollution in Rural Areas

In rural areas, air pollution is considered less serious as compared to water pollution(Bayram et al, 2001). Effects on health have been researched with various formations by different associations, and a connection between daily contamination ratio and general respiratory problem was determined(Bircan, 2003). Respiratory illnesses have been shown to increase in December-January in children as well as adults (Bosson et al., 2003; Devalia et al., 1994).Another study has found that there is positive relation between allergies and air pollution(Diaz-Sanchez, 1994). It is observed that there is a significant relation between particulate contaminants and SO_2 levels (Elbir, 2000). In another study, it is found that there has been significant connection between respiratory illnesses with SO_2 levels (Fişekçi et al., 1999). While yet another study reported that there is a linear connection between particulate contaminants and air's relative moisture(Fişekçi et al., 2000). Connections between air pollution and asthma have also been found (Harrod et al., 2003) as well as between SO_2 and particulate contaminants(Kaygusuz et al., 2003). There are other studies on connections between SO_2 and particulate contaminants(Keles et al., 1999). Besides, respiratory illnesses rising from asthma and SO_2 concentrations have increased (Koenig, 2016).

EFFECTS OF AIR POLLUTION ON HUMAN HEALTH: CAUSES OF AİR POLLUTİON

In laboratory studies regarding air contaminants, air contaminants have been shown to cause various respiratory illnesses such as asthma(Koike et al., 2004; Kumar et al., 2017). Studies of the effect of ozone on humans show that permeability of respiratory tracts increase on exposure to ozone, and there is an increase in the eosinophil concentration in the respiratory tract . It is reported that eosinophil concentrations increase in people suffering from asthma people (Koike et al., 2004). Another study reported that ozone causes inflammatory cytokine in lung bronchial mucosa(Lipfert, 1994). Additionally, ozone increases antigen server activity in rat lungs (Nemmar et al., 2003). Fewer studies have been performed with NO_2 than ozone, but the results were similar to ozone in people suffering from asthma (Koike et al., 2004). Studies on the combined effects of gasses, SO_2 and NO_2 show increases in allergic reactions in the respiratory system.(Özer et al., 1997).

Studies on diesel exhaust, an important component in particulate air pollution, shows that exposure to this gas increases concentrations of cytokine and inflammatory cells such as neutrophil and lymphocyte in liquids in respiratory tracts in humans (Panswad et al., 2016).In Experimental human studies performed with particles that are obtained from diesel exhaust, allergic reactions were shown to increase(Peden, 1997). From experimental studies on rats, it is reported that diesel exhaust particles cause respiratory illnesses(Phung et al., 2016). Besides, diesel exhaust particles increase respiratory illnesses in mice (Hua et al., 2016). To protect the respiratory system against virus, researchers have begun to investigate the role of air contaminants. Bayram et al., found that when cell particles obtained from healthy and asthmatic people were exposed to ozone, there was an increase in permeability in asthmatic cells (Qureshi el al., 2016). On the other hand, this gas increases the number of people suffering from asthma (Rusznak et al., 1997). In the same experimental conditions, similar results were obtained with NO_2.

Bayram et al., in studies performed with diesel exhaust particles, found that these particles exacerbate problems of respiratory system illness (Sango et al., 1997). These results show that effects of diesel exhaust particles on asthmatic people depend on their concentration(Sango et al., 1997). Besides, it is found that diesel exhaust particles increase proliferation of lung alveoli. They show that diesel exhaust particles accelerate the cell cycle. This effect of diesel exhaust particles happens to be pressured by N-asetyl sistein. Diesel exhaust particles repress expression of this protein (Sastry, 2002).

FIGHT AGAINST AIR POLLUTION

Global Air Pollution Mitigation Efforts

The studies for preventing air pollution in the world are held to standard air quality by enforcing laws, controlling emission, and analyzing the limiting negative effects of healthy people on public health. Control processes involve identifying emission sources, characterization of emissions, and monitoring of air pollution. Other initiatives include checks on sources of air pollution, filters, catalytic convertors, and improving combustion standards for emission control. WHO has prepared regulations for creating air quality standards. For this, acceptable maximum standard values have been created to protect the groups under risk determined short time particle concentration in air, and to protect populations from determined average annual particle concentrations. For applying these determined values, several regulations and laws have been prepared in efforts to maintain air pollution levels below limits set by WHO.

Air Pollution in Asia: Preventions and Available Case Study

Asia experienced rapid urbanization in the 1950s, with fast growth of population, and as a result, coil and petroleum requirements increased, and consequently, so did air pollution episodes (Sharma et al., 2016). Rampant urbanization, increases in motor vehicles, and inadequate combustion techniques, all contribute to air pollution (Stenfors et al., 2004). Air pollution that contains SO_2 and particulate matter is monitored by government regulation since 1985. In some stations that measure air pollution, NO_x and CO measures have been performed (Tam et al., 2016). But, in most Asian countries, O_3 and NO_2 measurements cannot be performed as we lack the necessary technology. Today's air pollution regulations originate from environmental regulations and laws made in 1983, with the primary goals of controlling emission of contaminants, and protecting humans from the negative effects of air pollution (Tam et al., 2016; West et al., 2016).

According to these regulations, by considering WHO and European Union Standards, SO_2 and particulate matter limitation were determined for short and long terms. Local authorities are responsible for enforcing these regulations when SO_2 and particulate materials exceed limit values (West et al., 2016).

Air pollution is mainly caused by industry, fuels and exhaust emission from traffic (Stenfors et al., 2004; Tam et al., 2016; WHO, 1992). In addition to limit emission from exhaust, burned fuel, industry, development of renewable clean energy sources and more efficient use of these are investigated. For this, controls for limitation emissions from industry, and some agreements for limiting emissions in cement and iron-steel industries have been signed(Tam et al., 2016). Use of fuel that contains low concentration of SO_2 and particulate material at home, use of higher coal grades, use of natural gas, heat isolation in buildings, educating heater personnel are some preventative measures for limiting air pollution(Tam et al., 2016). Checking of Exhaust emission, using fuels such as benzine without lead, limiting of exhaust emission, and encouraging vehicles that have catalytic converters are some implementations that improve air pollution in traffic. Researchers have identified the most effective renewable clean energy sources, as well as methods for increasing the capacity of hydroelectric energy, and more efficient methods to utilize solar energy and bioenergy (Tam et al., 2016). Methods to prevent migration to big cities, increasing green areas in cities, smart designing of cities, and educating the public regarding air pollution are other subjects in these studies (Sharma et al., 2016, Spengler et al., 1996).

Results from air pollution observations in Asian cities since the 1990s indicate a dramatically decreasing trend in air pollution levels, indicating that the measures taken to control air pollution have been effective. Besides, it is seen that concentrations of SO_2 and particulate matter in winter months decreased (Tam et al., 2016). Use of natural gas in buildings and industry, using of high quality coal, coal enrichment, advancement in combustion systems are effective in the abatement of air pollution (Tam et al., 2016). However, air pollution levels in rural Asia is increasing at an alarming rate (West et al., 2016). Therefore, when air pollution decreases in big cities, it increases in rural areas (West et al., 2016). This problem is created by inadequate controls on emissions of air pollution in rural Asian cities and use of uncontrolled low grade coal.

Air Pollution In Asian Cities, Current State and Precautions Taken

In recent years, people from many rural Asian cities have migrated to metropolitan cities. Along with this, unplanned urbanization and industrialization increase air pollution significantly. In Many Asian cities, air pollution measurements were performed in two different points, but these points do not represent whole cities. In December 2004, in addition to SO_2 and particulate matter measurements, NO_x and O_3 measurements were performed.

Considering the sources of air pollution, it is evident that the solid and liquid fuels used for heating in the winter months do not conform to government standards. Moreover, the sale and distribution of these fuels are often not adequately controlled Air pollution from traffic creates significant problems in Asian cities. To evaluate air pollution, on exhaust of vehicles, CO, NO_x, SO_2, HC emission were controlled, but these controls have proven to be inadequate. Air pollution from industry in Asian cities have increased significantly in recent years. For some rural Asian countries, air pollution problems have reached to significant levels but, due to financial limitations, required measurements cannot be implemented. These measurements were performed with mobile exhaust emission devices. Precautions taken in traffic seems limited to cities centers.

At the out of checking emission that is created by traffic and industry as precautions for limiting air pollution, heater operation times have been checked for limiting emissions due to heating by taking samples from fuels and checking for other solid and liquid fuels in the buildings. The sale of low quality and non-standard fuels have been stopped. Besides, regular checks are conducted in other buildings and precautions have been taken for non-standard fuels.

When air pollution in big Asian cities are considered, annual concentrations of SO_2 and particulate matter have been as high as 110 $\mu g/m^3$ in 2000-2001, this level has increased in 2002, and decreased somewhat in 2003. In January 2004, SO_2 and particulate matter values went up to 134 and 137 $\mu g/m^3$, in December and concentration of SO_2 went up 115 $\mu g/m^3$, concentration of particulate matter went up to 120 μg/m3. Therefore, these values are above levels recommended by WHO .

Precautions to Limit Air Pollution in Asian Countries

For monitoring air pollution effectively, conducting continued air pollution measurements by trained personnel is important. Use of fuel combustion should be controlled by local government representatives and associations. In buildings with heaters, regular cleaning of the heater stack and education of heater personnel is important. Standard contaminant concentration levels of solid fuel have not been determined by WHO. Due to intensive urbanization, to abate emission, for small rural area, standard parameters should be set in place. Additionally, the public should be educated on the health risks of low quality coal.

REFERENCES

Akimoto, H. (2017). *Overview of Policy Actions and Observational Data for PM2. 5 and O3 in Japan: A Study of Urban Air Quality Improvement in Asia*. Academic Press.

Barth, Monache, Ghude, Pfister, Naja Manish, & Brasseur. (2017). An Overview of Air Quality Modeling Activities in South Asia. In Air Pollution in Eastern Asia: An Integrated Perspective (pp. 27-47). Springer.

Bayram, H., Devalia, J. L., Khair, O. A., Abdelaziz, M. M., Sapsford, R. J., Sagai, M., & Davies, R. J. (1998). Comparison of ciliary activity and inflammatory mediator release from bronchial epithelial cells of nonatopic non-asthmatic subjects and atopic asthmatic patients and the effect of diesel exhaust particles in vitro. *The Journal of Allergy and Clinical Immunology*, *102*(5), 771–782. doi:10.1016/S0091-6749(98)70017-X PMID:9819294

Bayram, H., Ito, K., & Chung, K. F. (2006). Regulation of human lung epithelial cell numbers by diesel exhaust particles. *The European Respiratory Journal*, *27*(4), 705–713. doi:10.1183/09031936.06.00012805 PMID:16455839

Bayram, H., Khair, O. A., & Abdelaziz, M. (2002). Effect of ozone and nitrogen dioxide on the permeability of bronchial epithelial cell cultures of non-asthmatic and asthmatic subjects. *Clinical and Experimental Allergy*, *32*(9), 1285–1292. doi:10.1046/j.1365-2745.2002.01435.x PMID:12220465

Bayram, H., Sapsford, R. J., Abdelaziz, M., & Khair, O. A. (2001). Effect of ozone and nitrogen dioxide on the release of pro-inflammatory mediators from bronchial epithelial cells of non-atopic non-asthmatic subjects and atopic asthmatic patients, in vitro. *The Journal of Allergy and Clinical Immunology*, *107*(2), 287–294. doi:10.1067/mai.2001.111141 PMID:11174195

Bircan, H. A. (2003). Effects of atmospheric sulphur dioxide and particulate matter concentrations on emergency room admissions due to asthma in Ankara. *Tüberküloz ve Toraks Dergisi*, *51*, 231–238. PMID:15143399

Bosson, J., Stenfors, N. N., Bucht, A., Helleday, R., Pourazar, J., Holgate, S. T., & Blomberg, A. et al. (2003). Ozone-induced bronchial epithelial cytokine expression differs between healthy and asthmatic subjects. *Clinical and Experimental Allergy*, *33*(6), 777–782. doi:10.1046/j.1365-2222.2003.01662.x PMID:12801312

Devalia, J. L., Rusznak, C., Herdman, M. J., Trigg, C. J., Davies, R. J., & Tarraf, H. (1994). Effect of nitrogen dioxide and sulphur dioxide on airway response of mild asthmatic patients to allergen inhalation. *Lancet, 344*(8938), 1668–1671. doi:10.1016/S0140-6736(94)90458-8 PMID:7996960

Diaz-Sanchez, D. D. (1997). The role of diesel exhaust particles and their associated polyaromatic hydrocarbons in the induction of allergic airway disease. *Allergy, 52*(suppl 38), 52–56. doi:10.1111/j.1398-9995.1997.tb04871.x PMID:9208060

Elbir, T., Müezzinoğlu, A., & Bayram, A. (2000). Evaluation of some air pollution indicators in Turkey. *Environment International, 26*(1-2), 5–10. doi:10.1016/S0160-4120(00)00071-4 PMID:11345738

Fişekçi, F., Özkurt, S., & Başer, S. (1999). Effect of air pollution on COPD exacerbations. *The European Respiratory Journal, 14*(Suppl 30), 393s.

Fişekçi, F., Özkurt, S., & Başer, S. (2000). Air pollution and asthma attacks. *The European Respiratory Journal, 16*(Suppl 31), 290s.

Harrod, K., Jaramillo, R., Rosenberger, C., Wang, S.-Z., Berger, J. A., McDonald, J. D., & Reed, M. D. (2003). Increased susceptibility to RSV infection by exposure to inhaled diesel engine emissions. *American Journal of Respiratory Cell and Molecular Biology, 28*(4), 451–463. doi:10.1165/rcmb.2002-0100OC PMID:12654634

Hua, Q., & Liao, T. F. (2016). The association between rural–urban migration flows and urban air quality in China. *Regional Environmental Change, 16*(5), 1375–1387. doi:10.1007/s10113-015-0865-3

Kaygusuz, K., & Sarı, A. (2003). Renewable energy potential and utilization in Turkey. *Energy Conversion and Management, 44*(3), 459–478. doi:10.1016/S0196-8904(02)00061-4

Keles, N., Ilicali, C., & Deger, K. (1999). Impact of air pollution on prevalence of allergic rhinitis in Istanbul. *Archives of Environmental Health, 54*(1), 48–51. doi:10.1080/00039899909602236 PMID:10025416

Koenig, J. (2016). 22 Sulfur Dioxide Exposure in Humans. *Toxicology of the Nose and Upper Airways, 334.*

Koike, E., Watanabe, H., & Kobayashi, T. (2004). Exposure to ozone enhances antigen presenting activity concentration dependently in rats. *Toxicology, 197*(1), 37–46. doi:10.1016/j.tox.2003.12.007 PMID:15003332

Lipfert, F. (1994). *Air Pollution and Community Health- A Critical Review and Data Sourcebook*. Van Nostrand Reinhold.

Nemmar, A., Hoet, P. H. M., Dinsdale, D., Vermylen, J., & Hoylaerts, M. F. (2003). Diesel exhaust particles in lung acutely enhance experimental peripheral thrombosis. *Circulation*, *107*(8), 1202–1208. doi:10.1161/01.CIR.0000053568.13058.67 PMID:12615802

Özer, U., Aydin, R., & Akçay, H. (1997). Air pollution profile of Turkey. *Chemistry International*, *19*, 190–191.

Panswad, T., Polprasert, C., & Yamamoto, K. (Eds.). (2016). *Water Pollution Control in Asia: Proceeding of Second IAWPRC Asian Conference on Water Pollution Control Held in Bangkok, Thailand*. Elsevier.

Peden, D. (1997). Mechanisms of pollution induced airway disease: In vivo studies. *Allergy*, *52*(suppl 38), 37–44. doi:10.1111/j.1398-9995.1997.tb04869.x PMID:9208058

Phung, D. T., Hien, T. T., Linh, H. N., Luong, L. M. T., Morawska, L., Chu, C., & Thai, P. K. et al. (2016). Air pollution and risk of respiratory and cardiovascular hospitalizations in the most populous city in Vietnam. *The Science of the Total Environment*, *557*, 322–330. doi:10.1016/j.scitotenv.2016.03.070 PMID:27016680

Qureshi, M. I., Awan, U., Arshad, Z., Rasli, A. M., Zaman, K., & Khan, F. (2016). Dynamic linkages among energy consumption, air pollution, greenhouse gas emissions and agricultural production in Pakistan: Sustainable agriculture key to policy success. *Natural Hazards*, *84*(1), 367–381. doi:10.1007/s11069-016-2423-9

Rusznak, C., Devalia, J. L., Bayram, H., & Davies, R. J. (1997). Impact of the environment on allergic lung diseases. *Clinical and Experimental Allergy*, *27*(s1), 26–35. doi:10.1111/j.1365-2222.1997.tb01823.x PMID:9179442

Sango, H. A., Testa, J., Meda, N., Contrand, B., Traoré, M. S., Staccini, P., & Lagarde, E. (2016). Mortality and morbidity of urban road traffic crashes in Africa: Capture-recapture estimates in Bamako, Mali. *PLoS One*, *11*(2), e0149070. doi:10.1371/journal.pone.0149070 PMID:26871569

Sastry, N. (2002). Forest fires, air pollution, and mortality in southeast Asia. *Demography*, *39*(1), 1–23. doi:10.1353/dem.2002.0009 PMID:11852832

Sharma, Ojha, Pozzer, Mar, Beig, Lelieveld, & Gunthe. (2016). *WRF-Chem simulated surface ozone over South Asia during the pre-monsoon: Effects of emission inventories and chemical 2 mechanisms 3*. Academic Press.

Spengler, J. D., & Wilson, R. (1996). *Particles in Our Air, Concentrations and Health Effects*. Harvard University Press.

Stenfors, N., Nordenhall, C., Salvi, S., Mudway, I., Soderberg, M., Blomberg, A., & Sandstrom, T. et al. (2004). Different airway inflammatory responses in asthmatic and healthy humans exposed to diesel. *The European Respiratory Journal*, *23*(1), 82–86. doi:10.1183/09031936.03.00004603 PMID:14738236

West, J., Cohen, A., Dentener, F., Brunekreef, B., Zhu, T., Armstrong, B., & Wiedinmyer, C. et al. (2016). What we breathe impacts our health: Improving understanding of the link between air pollution and health. *Environmental Science & Technology*, *50*(10), 4895–4904. doi:10.1021/acs.est.5b03827 PMID:27010639

WHO-United Nations Environment Programme. (1992). *Urban Air Pollution in Megacities of the World*. Blackwell.

Chapter 3
Assessment of the Chemical Precipitation Process as a New Approach for Industrial Emission Abatement Systems:
Assessment of the Chemical Precipitation Process

Ahmet Ozan Gezerman
Yildiz Technical University, Turkey

Burcu Didem Çorbacıoğlu
Yildiz Technical University, Turkey

ABSTRACT

The abatement of emission gases, such as SOx, NOx, and COx, is one of the main problems studied by researchers for continuous developments, necessitating considerable investments by several industries. Currently, the scrubber system with its use form, and the chemical precipitation method that is considered as an alternative, are the two different processes that have demonstrated the best results for emission abatement. In this chapter, an assessment is performed on an industrial scale for both the processes, their comparative advantages are discussed, and possible applications presented.

DOI: 10.4018/978-1-5225-3379-5.ch003

INTRODUCTION

Currently, in several industries, there are environmental regulations to control emission and various preventive measures have been realized according to the process structure. As solutions to the emission problem, till date, several processes have been developed. For emission abatement, the ease of use of the proposed process, and the energy and expense entailed are most important topics that require attention. Therefore, most proposed emission abatement methods in the industry currently, are mist eliminator processes that contain scrubbers. In a mist eliminator system, the emission gases in the source are sent to multiperforated layers after they condensed to known density values. Over the last 20 years, mist eliminators have been implemented for emission gas abatement in the petroleum and fertilizer industries. There are several examples of mist eliminator implementations that are used for COx, NOx, and SOx abatement. The removal of sulphur dioxide from the reaction ambience using mist eliminators is an important abatement process1. For the elimination of emission gases, such as SOx, NOx, and COx, several processes have been developed and in recent years, process developments have gained momentum (Wesley et al., 2000).

Another process that has been developed for mist elimination is a frusto-conical shaped protrusion system. The emission abatement efficiency in such a layered mechanical system is cost effective and has minimum chemical wastes (Jarrier et al., 2012). A prilling system that separates gases and liquids using mist a separator (Parks, 2001) is also available. Another mechanical system developed for mist elimination is wave-plated and has been implemented in the petroleum and fertilizer production industry (Azzopardi et al., 2002). Elimination methods that utilize the precipitation of fine particles in the emission gases have been developed (Altman et al., 2001). There are several processes in which the exact values of the air pollution levels are detected but they have some problems in terms of the process feasibility,including cost problems(Boyden et al., 2006). To prevent the contamination of the emission gasses, a combined coalescing media and a centrifugal cyclone system are highly beneficial in industrial production (Birmingham et al., 2001). To abate emission waste, processes that work at different pressure and temperature conditions are available and their implementation have been successful (Austrheim et al., 2008). There are some systems that minimize NOx emission using acid spray wet scrubbers. Such a process has been used to reduce emission gas wastes such as ammonia (Hadlocon et al., 2014). In another abatement system, gases emitted to the air have been removed using a condensing heat exchanger (Bielawski, et al., 2001). Systems using venturi scrubbers and absorbers for the elimination of emission gaseshave been most advantageous for the industry (Hargrove et al., 2004) . An example of the mist eliminator used for minimizing emission is the multilayered eliminator

(Nieuwoudt, 2011). Another eliminator system is a process that has a flooded-bed dust scrubber. In this process, good results were obtained for emission abatement on an industrial scale (Goodman, 2000). One of the processes for minimizing the harmful environmental effects of the emission gases in nitrogenous fertilizer is a full-scale spray scrubber. İn this process, which has a wide application area, the recovery of ammonia is possible (Hadlocon, et al., 2015). For addressing the emission problem during the production of acids such as sulphuric acid and nitric acid, similar processes have been developed (Diamond, 2011). In acid production, reactions between the nitrogen oxides and ozone have been developed in industrial applications for removing the emission sourced from nitrogen oxide (Suchak, et al.,2009). Significant mechanical systems developed for solving the emission problem can be used in the multi-layered elimination process. With this configuration, the emission problem can be significantly reduced, gradually (Kanka et al., 2012). Wet scrubbers are currently being used for decreasing industrial air pollution. Several studies have been undertaken for evaluating the performances of these processes (Danzomo et al., 2012). Assessments of these systems have been most effective for industrial applications with respect to the production costs (Sorensen et al., 2000). The performance of the scrubber process was found to be more effective at a higher pressure (Setekleiv el al, 2010). Another similar study on the abatement of emission gases in industrial processes is the water absorption by the purging method (Gezerman, 2015; Gezerman, 2016). Despite its feasibility for industrial applications, this process does not have a high contamination like the emission systems and does not need a scrubber system; it can be applied to chemical precipitation systems easily.

In this study, the differences and similarities between the mist elimination performed by the chemical precipitation method and a mist eliminator with a scrubber are discussed. Additionally, an alternative is presented for ease of industrial use.

MIST ELIMINATORS FOR EMISSION ABATEMENT

In order to work at various temperatures in different processes, mist eliminators in industrial facilities (Figure 1) are required to be constructed using materials that have physical and mechanical resistance. These materials must be selected for maximum effectivess in muti-gradual operations to ensure that the required results are obtained for the removal of the waste gases (Figure 2). Till date, in mist eliminator processes, the optimum operation temperature is between -15–204°C. In order to increase the efficiency of the eliminator by increasingthe number of layers, various types and forms of layers can be produced and utilized.

Recently, it has been established that that the geometrical shape and structure of each layerhas an effect on the efficiency of the eliminator, except for the multilayer

Figure 1. Eliminator system containing a one-layered scrubber(Michelson, 1986)

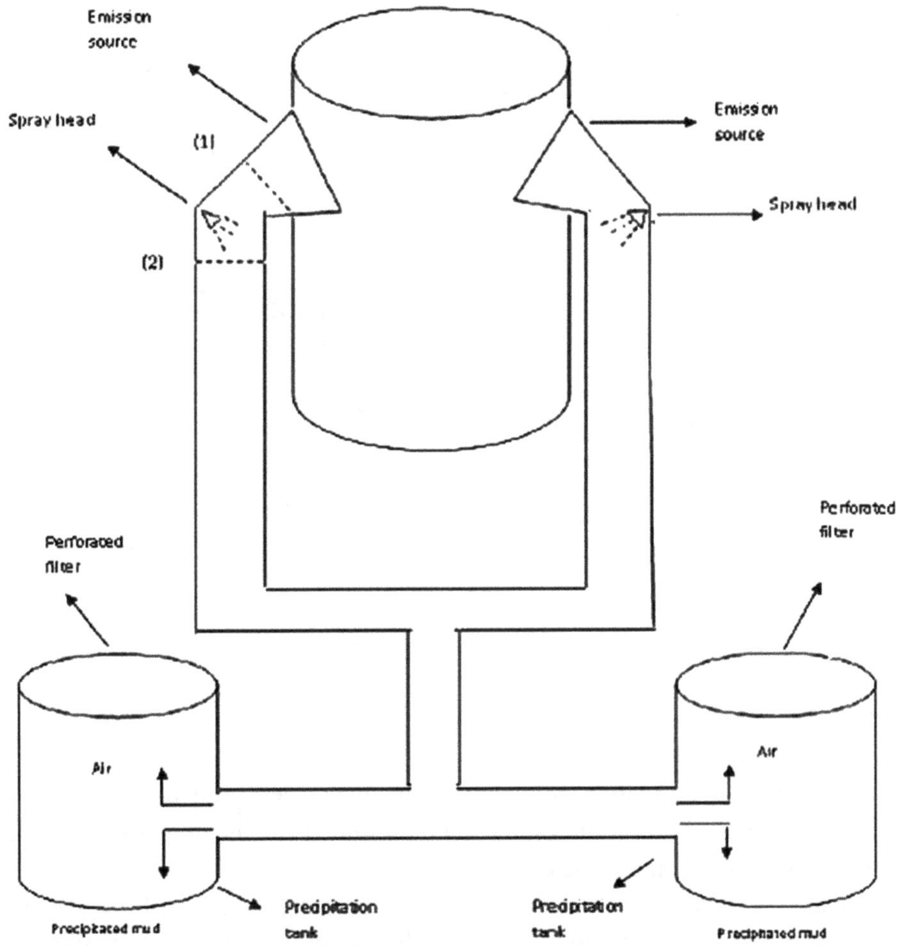

structure of the mist eliminator for emission abatement. Similar to the number of layers, the layer material, the holes in these layers and diameters of these holes increase the abatement efficiency.

Mist eliminators that are used in the industry can be designed for specific industrial processes in certain cases. In acid and fertilizer production, owing to the maximum emission of SOx and NOx and the requirement to condense the emission gasses, it is proposed to attain a liquid phase using a water spray (Figure 3). Therefore, the waste gas is retained before emission to the atmosphere and it is aimed to condense it further. As in this design, the number of holes on the perforated plates' surface is increased and each plate is made of a different material with a different condensation temperature to ensure a maximum condensation of the gas into a liquid state.

Figure 2. Multilayer scrubber mist eliminator system (VanBuskirk el al., 1995)

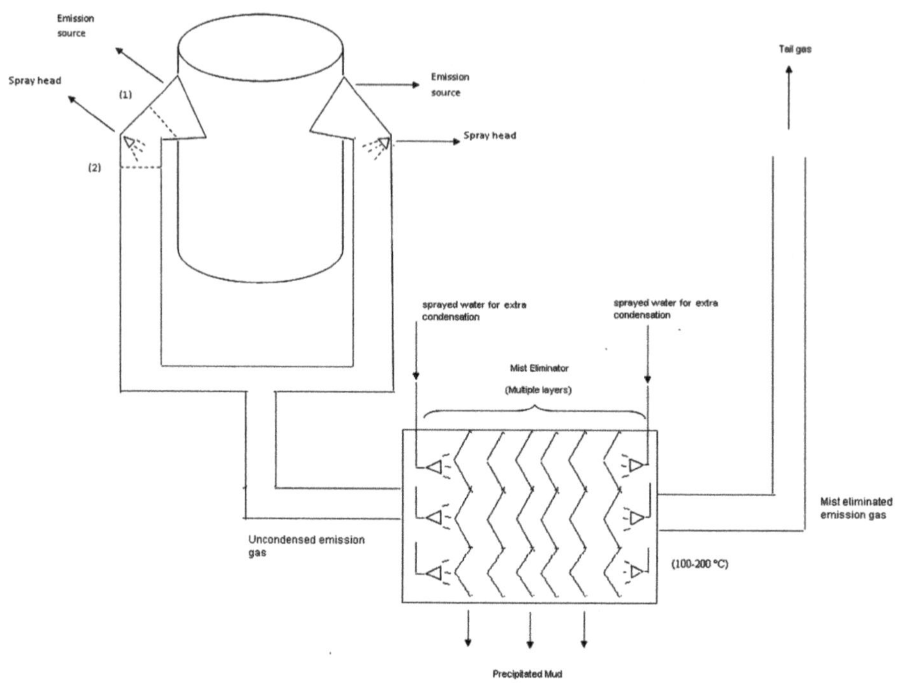

The chemical composition of the fiber layers is designed to work between 100–250°C because the final product can be produced in this temperature range in fertilizer production technologies.

The usage of mist eliminators in emission abatement systems have started to increase. Initially, mist eliminators with scrubbers contained a single retention layer with a single type of material. Further, to reduce SOx and NOx in the emission gases, the number of layers was increased but only one type material was used. Currently, the number of layers has been increased, and the type of material used for each layer plate has been changed, aiming to increase the emission abatement efficiency.

In mist eliminator operations, reactions between the liquid and gaseous phases result in small droplets named aerosols. This operation can be described as mist removal or a condensation of the waste gas to liquid, mechanically. These eliminators are designed to have large surface areas for small particle volumes to collect the liquid droplets.

The elimination of waste gas in the mist eliminator is realized using mechanical mist separators. The regular functioning of the gas outlet is to be ensured; else, pressure loss problems can be created in the process. The purpose of using mechanical

Figure 3. Emission abatement system with a multilayer scrubber for mist elimination

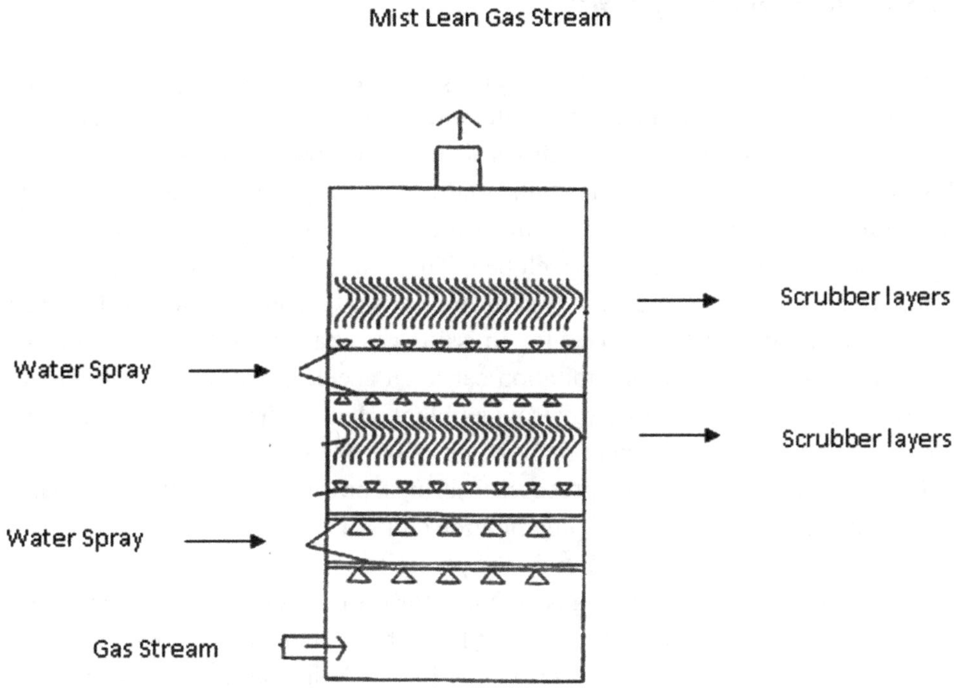

separators is to condense the gas into a liquid by retaining the gas, which crashes onto the plate and to ensure a sufficiently large surface area for the gas to condense.

The formation of small mist particles is inevitable in the manufacturing process. These particles cause equipment contamination, corrosion, contamination in the heat exchange, and other damages. When these mist particles are released into the atmosphere, they cause air pollution, violating the emission standards.

Elimination of waste gasses can be performed in various methods in industrial facilities. Generally, three processes are used. In two of these, mist having small particles is created by chemical reactions and large particles from the condensation can be atomized using mechanical equipment. The other method is the condensation of the waste gases by cooling. Finally, to obtain a higher density mist, the temperature and pressure can be varied in the process.

Currently, the features preferred by producers in several industrial facilities for mist eliminator systems include ease of use for industrial processes, impurity abatement, low pressure applications, usage in industrial applications with high contamination, low equipment maintenance, and high efficiency.

CHEMICAL PRECIPITATION METHOD FOR EMISSION ABATEMENT

There are few applications in the industry for gas treatment by the chemical precipitation method. As facilities containing steam turbine generators meet the energy requirements of the factory, this case does not cause additional energy loads. However, energy consumption is a vital and important parameter in cost analysis. Therefore, for an emission gas abatement process, a system that does not require extra energy costs needs to be developed. The chemical precipitation process is similar to a heavy metal removal system (Figure 4). Chemicals used in the heavy metal removal process can be used for emission gas abatement. Besides, the mixer equipment that is used for precipitation cause an acceleration in the reaction and the reaction's pressure moves away from ambient. Hence, chemical precipitation is more effective for gas abatement than the mist- scrubber separators.

In the chemical precipitation process, differing from the mist scrubber eliminator, there are perforated filters on the precipitation tanks. These perforated filters ensure the emission of the gases generated in the precipitation reaction.

Table 1 depicts the emission gas amounts for the emission gases emitted in the prilling tower. All the calculations are for 150.000 m^3/h emission gases.

It is assumed that the factory, where the fertilizer is being produced, has a capacity of 100 t/h and has cooler motors with a power rating of 40 kW for the prilling and granulation processes; the volume of air in the cooling tower cone is 150.000 m3. Then, for a typical fertilizer production process, as per the probable reaction of calcium ammonium nitrate, the rate at which the gases are emitted during prilling will be 150.000 m^3/h. The mixture of the emission gases (1) is listed in Table 1.

The average molecular weights of the emission gases can be calculated using these percentage values. These calculations will aid in determining the vapor pressure of the gases exiting the cooling towers.

The molecular weight of air is $(0.79 \times 28) + (0.21 \times 32) = 22.12 + 6.72 = 28.84$ g/mol. The molecular weight of the emission gases at the selected point is given by,

$(18 \times 2\%) + (80 \times 0.08\%) + (132 \times 0.02\%) + (102 \times 0.0000025\%) + (100 \times 0.0000025\%) +$

$28.84 \times 0.979 = 28.23436 + 0.4504 = 28.68476$ g/mol.

As per the ideal gas equation, $P \times MW = d \times R \times T$,

where P is the gas pressure, MW is the molecular weight, d is the density of the gas emitted, R is the universal gas constant, and T is the temperature.

$P = 1.145 \times 0.082 \times (273+50)/28.68476 = 1.057$ atm.

In these emission gases, when water is added over 1 h to condense the 50% water present in the emission gases,

$$m = 2.08717 \times 10 - 2 - 2.21315 \times 10\text{-}4\ Tw + 2.0416 \times \sqrt{\theta}\ (e\text{-}4217.96/Tw) + mo$$

where mo is the initial amount of emission gas at the outlet of the prilling tower (1), Tw is the dewpoint of water at the spray head, m is the amount of water after condensation at the final point (2), θ is the diffusion coefficient of the emission gases at 50 °C (for this particular mixture of emission gases).

Using this formula, we can determine the amount of water present in the emission gas mixture. The same calculation can be made for the emission gases at 50 °C (Tw). The result obtained shows that given the stoichiometric ratios of the condensed gas, the mass of the emitted gases is 44% of the initial mass. After the condensation of the emission gases, which are 44% of the initial mass, the materials remaining in the emission gases are as given below. After condensation (2), the emission of air at a rate of 150.000 m^3/h and the emission gas mixture are given in Table 2.

After this selected point (2), the molecular weight of the emission gases that condense and are 44% of the initial mass are as follows:

$(18 \times 0.88\%) + (80 \times 0.0352\%) + (132 \times 0.0088\%) + (102 \times 0.0000011\%) + (100 \times 0.0000011\%) + 28.84 \times 0.99076 = 28.77169$ g/mol.
$P_2 \times MW = d \times R \times T$
$P_2 = 1.145 \times 0.082 \times (273 + 35)/28.77169 = 1.005 =$ approximately 1 atm.

Because the pressure, P_1, is greater than the pressure, P_2, the value of P_2 will not influence the value of P_1 at the exit of the prilling tower.With these pressure values, after the abatement of the emission gases after condensation, the solidification of harmful chemicals will be possible by chemical precipitation. Generally, these methods are used for the removal of heavy metals. The same reactions occur during these removal methods, as well. These methods are based on the differences in the solubilities of chemicals at different pH levels. The notable reactions are as follows (Gezerman, 2012):

Reaction for the precipitation of ammonium sulfate by FeC13:

$$3(NH_4)_2SO_4 + 2FeCl_3 \rightarrow 6NH_4Cl + Fe_2(SO_4)_3$$

Table 1. Mixture of the emission gases

Emitted gas	Rate (m^3/h)	Percentage in the mixture (%)
H_2O	3,000	2
$(NH_4)_2SO_4$	30	0.02
NH_4NO_3	120	0.08
$Ca(NO_3)_2$	0.00375	0.0000025
$CaCO_3$	0.00375	0.0000025

Reaction for the precipitation of ammonium nitrate (with a pH value of 9) by adding NH_4OH:

$$NH_4NO_3 + NH_4OH \rightarrow NH_3 \text{ (gas)} + NH_4NO_3 + H_2O$$

Reaction for the precipitation of $Ca(NO_3)_2$ (with a pH value of 12) by adding NH_4OH:

$$Ca(NO_3)_2 + NH_4OH \rightarrow NH_3 \text{ (gas)} + Ca(NO_3)_2 + H_2O$$

RESULTS AND DISCUSSION

These emission abatementsystems have various differences between them. These systems, according to the emission gas compositions, can contain specific properties. In industrial processes, till date, the method that is most widely used is the mist eliminator that contains scrubbers. These methods have several advantages and disadvantages as emission abatement systems. Initially, when developed, mist eliminator systems with scrubbers having one type of material and one layer were used. These systems operated at a constant temperature. For enhancing the performance, multilayered scrubber systems were considered for mist eliminators. Accordingly, it was aimed to obtain the best performance by increasing the retention time in the mist eliminator; these changes increased the process performance to 30%. On the other hand, differences in the raw materials in mist eliminators with multilayered scrubbers are significant for emission abatement. Here, each plate should have a polymeric structure and the operation temperature should be between 100–200 °C. Therefore, in the prilling tower, the spray temperature changes between 100–200 °C. These multilayered, efficient emission abatement systems containing scrubbers in which each plate is made from a different material, are 20% larger than the other previous emission abatement systems. Further, the total abatement capacity can go up to 70%.

As a different process in mist eliminator systems, the waste emission gases are condensed andsent to a chemical precipitation tank for obtaining the chemical mud. This method is similar to a mist separator that contains scrubbers. However, in the chemical precipitation method, different from the mist separators containing scrubbers, there are no layered scrubbing plates that are used for increasing the retention time.Instead, this performance is expected from the chemical precipitation process. Moreover, in a layered scrubber system, for preventingpressure changes in the emission abatement process, a fan for suction purposes is placed on the outlet of the scrubber. As the scrubber plate layers cause pressure drops owing to long retention times, the suction fan used in the industrial facilities poses considerable problems because of its energy consumption. Because fans are used in prilling for dehydration, the energy load of the process increases, as the consumed energy of the added fan increases.

Figure 4. Emission abatement system with a chemical precipitation method for the prilling tower

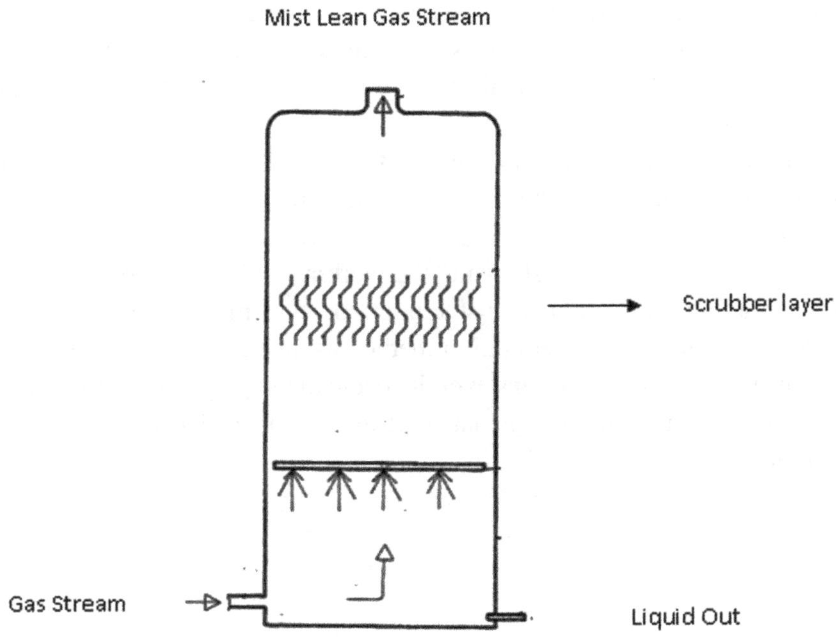

On the other hand, in industrial processes, the cleaning of the scrubber plates in mist separators posesignificant problems for the process continuity. This is because at eachcleanup process, the mist eliminator system mustbe shut down and the emission gases are sent into the atmosphere. In a chemical precipitation process, pressure changes do not occur as in the scrubber processes and the precipitation operation is conducted in more than one precipitation tank (Figure 4). Therefore, in the chemical precipitation method, during a maintenance period, the waste emission gases are not sent into the atmosphere.

Another advantage of the chemical precipitation method is that the chemicals, which are used for precipitation, are not different from the other chemical treatment materials such as acids and the other electrolyte solutions used in chemical industries. Moreover, the fan motor used in the mist eliminator system for preventing pressure dropsis not used in the chemical precipitation method because pressure drops do not occur and the energy cost of the chemical precipitation method is superior than that of the scrubber-mist eliminator system.

CONCLUSION

In this study, two elimination systems were compared: the scrubber-mist eliminator and the chemical precipitation process. The comparative advantages of these systems were discussed. Additionally, the emission abatement efficiency and industrial costs were also discussed. Accordingto this, the energy requirement of the mist eliminator system (>90 kW/h) creates extra costs.

With respect to the emission abatement efficiency, although the scrubber plates increase the retention time of the emission gases, the life of the scrubber plates are shortened because ofcorrosion. Therefore, the required time for cleaning the scrubber plates in the mist eliminator system and the workmanship increase.

Thus, although the scrubber-mist eliminator system has certain advantages compared to the chemical precipitation method as an emission abatement system, considering the corrosion life, the gases emitted into the atmosphere, and workmanship needed for the cleaning and maintenance of the scrubber plates, these advantages are not significant.

REFERENCES

Altman, R., Buckley, W., & Ray, I. (2001). Wet Electrostatic Precipitation Demonstrating Promise for Fine Particulate Control-Part II. *Power Engineering*, *105*(2), 42–42.

Austrheim, T., Gjertsen, L. H., & Hoffmann, A. C. (2008). Hoffmann, Alex C. "An experimental investigation of scrubber internals at conditions of low pressure. *Chemical Engineering Journal*, *138*(1), 95–102. doi:10.1016/j.cej.2007.05.048

Azzopardi, B., & Sanaullah, K. (2002). Re-entrainment in wave-plate mist eliminators. *Chemical Engineering Science*, *57*(17), 3557–3563. doi:10.1016/S0009-2509(02)00270-1

Bielawski, G. T., & Bhat, P. A. (2001). *Mist elimination/air toxic control in a wet scrubber using a condensing heat exchanger*. U.S. Patent No. 6,273,940. 14 Aug. 2001.

Birmingham, D. P., & Rush, G. C. (2001). *High efficiency gas scrubber using combined coalescing media and centrifugal cyclone*. U.S. Patent No. 6,251,168.

Boyden, S. A., & Piche, S. (2006). *Control of rolling or moving average values of air pollution control emissions to a desired value*. U.S. Patent No. 7,113,835.

Danzomo, B. A., Salami, M.-J. E., Jibrin, S., & Nor, I. (2012). Performance evaluation of wet scrubber system for industrial air pollution control. *Journal of Engineering and Applied Sciences (Asian Research Publishing Network)*, *7*(12), 1669–1677.

Diamond B. W. Removal of acid mists. (2011). U.S. Patent No. 8,025,860.

Gezerman, A. O. (2012). A practical solution to abate emission of ammonium nitrate and ammonia during prilling and granulation processes. *International Journal of Medicinal Chemistry*, *2*(2), 57–63.

Gezerman, A. O. (2015). Exergy analysis and purging of an ammonia storage system. *International Journal of Exergy*, *17*(3), 335–351. doi:10.1504/IJEX.2015.070502

Gezerman, A. O. (2016). Industrial-scale purging of ammonia by using nitrogen before environmental discharge. *International Journal of Industrial Chemistry*, *7*(4), 1–8. doi:10.1007/s40090-016-0096-6

Goodman, G. V. R. (2000). Using water sprays to improve performance of a flooded-bed dust scrubber. *Applied Occupational and Environmental Hygiene*, *15*(7), 550–560. doi:10.1080/10473220050028376 PMID:10893791

Hadlocon, L. J. S., Manuzon, R. B., & Zhao, L. (2014). Optimization of ammonia absorption using acid spray wet scrubbers. *Transactions of the ASABE*, *57*(2), 647–659.

Hadlocon, L. J. S., Manuzon, R. B., & Zhao, L. (2015). Development and evaluation of a full-scale spray scrubber for ammonia recovery and production of nitrogen fertilizer at poultry facilities. *Environmental Technology*, *36*(4), 405–416. doi:10.1080/09593330.2014.950346 PMID:25518983

Hargrove, O. W., & Denlinger, M. A. (2004). *Gas distribution system for venturi scrubbers and absorbers*. U.S. Patent No. 6,808,166.

Kanka, H., Jansen, M., Krauss, R., & Kaiser, R. (2012). *U.S. Patent No. 8,328,918*. Washington, DC: U.S. Patent and Trademark Office.

Michelson, I. (1986). *Chevron-type mist eliminator and method*. U.S. Patent No. 4,601,731.

Nieuwoudt, I. (2011). *Two-stage mist eliminator and method*. U.S. Patent No. 7,905,937.

Parks, C. R. (2001). *U.S. Patent No. 6,190,438*. Washington, DC: U.S. Patent and Trademark Office.

Rene, E., Jarrier, D. J., & Bryant, P. (2012). *U.S. Patent No. 8,273,158*. Washington, DC: U.S. Patent and Trademark Office.

Setekleiv, E., & Svendsen, H. (2010, January). Scrubber performance: Investigation of liquid holdup in mesh pad at ambient conditions with applications to high pressure separation. *Offshore Technology Conference*. doi:10.4043/20390-MS

Sorensen, I. W., & Lamb, E. A. (2000). *U.S. Patent No. 6,068,686*. Washington, DC: U.S. Patent and Trademark Office.

Suchak, N., Finley, S. J., Eschbach, J. A., & Aeiss, R. (2009). *U.S. Patent No. 7,632,475*. Washington, DC: U.S. Patent and Trademark Office.

VanBuskirk, G. K., Hsieh, C.-L., McNulty, K. J., & Hansen, O. V. (1995). *U.S. Patent No. 5,464,459.* Washington, DC: U.S. Patent and Trademark Office.

Wesley, B., Anderson, D. K., & Kingston, W. H. (2000). *U.S. Patent No. 6,083,302*. Washington, DC: U.S. Patent and Trademark Office.

Chapter 4
Global Warming and Climate Change:
Challenges and Impacts

Kijpokin Kasemsap
Suan Sunandha Rajabhat University, Thailand

ABSTRACT

This chapter indicates the advanced issues of global warming, the significant aspects of climate change, climate change and crop production, climate change and risk issues, and the public perceptions of climate change. Global warming and climate change are the important environmental concerns that have a major impact on various environmental perspectives. The effects and impacts of global warming are climate change, sea level change, water balance, and human health. The problems of global warming and climate change can be controlled by minimizing the emission of greenhouse gases into the environment through conducting laws, reducing thermal power generating stations, planting trees, and sharing vehicles. Global warming and climate change must be recognized by local governments and environment-related international organizations in order to find the effective approaches to reducing their impacts in environmental settings.

INTRODUCTION

Global warming, which is caused by the combustion of fossil fuels as the exhaustible resources (Winter, 2014) during economic development in human society (Reynolds, Bostrom, Read, & Morgan, 2010), threatens global ecological balance, imperils ecological safety, and endangers human race (Li & Wang, 2012). Emissions of

DOI: 10.4018/978-1-5225-3379-5.ch004

greenhouse gases due to human activities (e.g., extensive use of fossil fuels and deforestations) have altered greenhouse gases in the atmosphere in ways that are expected to generate global warming and climate change (Mukherjee & Mukerjee, 2017). The analysis of ongoing global temperature change is subject to the increasing scrutiny and criticisms that are different than would occur for a purely scientific problem (Hansen, Ruedy, Sato, & Lo, 2010).

The increasing level of concentration of greenhouse gases in the atmosphere has been directly linked to the changes being experienced in the world's climate (Bekele & Ganpat, 2017). Among the greenhouse gases, it is known that carbon dioxide (CO_2) is the major contributor to global warming (Goyal & Royal, 2015). Climate change is considered as a key threat to biodiversity and ecosystem functions (Graham & Harrod, 2009), affects human societies and environment (Pappis, 2011), results in a rapid expansion of shrubs in the Arctic (Kaarlejärvi, Hoset, & Olofsson, 2015), and relates to precipitation patterns (Heino, Virkkala, & Toivonen, 2009). Companies need to engage in give-and-take exchange relationships with other companies to address climate change (Finke, Gilchrist, & Mouzas, 2016).

Investigation is required to understand climate change-related risks for all components of human systems, including cultural heritage (Forino, MacKee, & von Meding, 2016). Revisions to the design of existing mitigation measures within existing policy frameworks are considered the most effective way to account for the impacts of climate change in the future mitigation planning (Mullan, Vandaele, Boardman, Meneely, & Crossley, 2016). In order to mitigate the effects of climate change, a lot of research has reduced CO_2 emissions and developed the capture and storage technologies (CCS) to reduce climate change crisis (de Richter, Ming, Caillol, & Liu, 2016), regarding the decrease in CO_2 emission and the progress of alternative energy sources (Popescu & Nica, 2017). The balance between ecosystem CO_2 uptake and release can determine the level of carbon sequestration in terrestrial ecosystems and its potential impact on CO_2 concentration in the atmosphere (Li et al., 2017).

This chapter focuses on the literature review through a thorough literature consolidation of global warming and climate change. The extensive literature of global warming and climate change provides a contribution to practitioners and researchers by indicating the challenges and impacts of global warming and climate change in order to increase the global environmental performance.

Background

Global warming has become one of the major challenges in maintaining global food security (Adhikari, Nejadhashemi, & Woznicki, 2015) and has influenced crop growth and water utilization (Guoju et al., 2016). The warmest world, being wealthier, should have greater capacity to address any problem, including global warming (Goklany,

2012). Beliefs regarding global warming are shaped by individual experiences and weather phenomenon (Borick & Rabe, 2010) and associated with the support for policies aiming at reducing global warming (Zhao, Leiserowitz, Maibach, & Roser-Renouf, 2011). The higher temperatures and the changes in seasonal precipitation patterns have consequences for the availability of water resources, soil quality, and diseases outbreaks, leading to the significant changes in the vegetative cycle, crop yields, and livestock productivity (Popović & Mijajlović, 2014).

Informing people about climate science can play an important role in mobilizing action to appropriately mitigate climate change (Ranney & Clark, 2016). Economic conditions are highly influential in determining whether climate change is covered and how the issue is framed (Boussalis, Coan, & Poberezhskaya, 2016). Evaluation of the ongoing efforts for farm level adaptation to climate change is crucial to understand their effectiveness and to suggest further actions at the policy level (Abid, Schneider, & Scheffran, 2016). Addressing specific environmental challenges requires the economic-technical solutions and requires addressing the social-political dimensions (Ploberger, 2013).

Climate change dominates both political and business areas (Elijido-Ten, 2017) recognized by scientists (Pearce, Brown, Nerlich, & Koteyko, 2015), policymakers (Schmeida & McNeal, 2017), and many industries, such as tourism (Grillakis, Koutroulis, Seiradakis, & Tsanis, 2016). In coastal destinations regarding tourism, climate change adaptation is needed to address the coastal erosion due to a combination of sea level rise and more frequent extreme weather events leading to the loss of natural features and tourism infrastructure (Schliephack & Dickinson, 2017).

IMPORTANT PERSPECTIVES ON GLOBAL WARMING AND CLIMATE CHANGE

This section explains the advanced issues of global warming; the significant aspects of climate change; climate change and crop production; climate change and risk issues; and the public perceptions of climate change.

Advanced Issues of Global Warming

The problem of identifying key economic sectors that significantly contribute to global warming is posed in mathematical terms as a bi-criteria linear program that seeks to simultaneously optimize the total economic output and the total life cycle CO_2 emissions (Cortés-Borda et al., 2015). The hiatus in global warming is possibly due to strong cooling in the tropical Pacific (Arora et al., 2016). Reducing greenhouse gas emissions and developing strategies to adapt to the future impacts of climate change are among the most pressing needs of the world today (Danos & Boulouta, 2011).

To avoid and reduce the intensity and severity of global warming and climate change, its mitigation is essential (Srivastava, 2017). If the emission of greenhouse gases is to be reduced, consumers need to be knowledgeable about global warming causes and consequences in order to adopt the innovation and participate in mitigation measures (dos Santos, 2012). Higher yields and lower greenhouse gas intensities and carbon costs can be achieved by substituting chemical nitrogen fertilizers with organic fertilization strategies (Zhang, Li, & Xiong, 2016).

Global warming increases nutrient pollution from anthropological sources (Chaves, Melchers, Peng, & Stewart, 2016) and causes a persistent shift from heterotrophic to more autotrophic control of the growing season carbon cycle in the carbon-rich permafrost ecosystems (Pries et al., 2015). The thaw and release of carbon frozen in permafrost can predict global climate change (Li et al., 2014) and can enlarge surface warming to initiate a permafrost carbon feedback (PCF) on climate (Schaefer, Zhang, Bruhwiler, & Barrett, 2011). Using seasonal insulation permits an incremental response to future climate warming conditions as they occur (Perreault & Shur, 2016).

A warmer climate leads to the decreased surface snow sublimation and the fraction of snowfall eroded and transported by the blowing snow (López-Moreno, Boike, Sanchez-Lorenzo, & Pomeroy, 2016). Global warming can influence the Atlantic multidecadal oscillation (AMO) variability, and the AMO can affect the sea surface temperature (SST) variability over the global ocean (Wang & Dong, 2010). Greening and the highly reflective materials have a positive impact on the mitigation of global warming and urban heat island (Yumino, Uchida, Sasaki, Kobayashi, & Mochida, 2015). Due to urban heat island and global warming, the higher temperatures have a serious impact on the electricity consumption of the building sector, thus increasing the peak and the total electricity demand (Santamouris, Cartalis, Synnefa, & Kolokotsa, 2015).

Global warming reduces both flower and seed production for herbaceous species (Liu, Mu, Niklas, Li, & Sun, 2012). The growth of terrestrial plant species in response to global warming determines the future dynamics of terrestrial vegetation (Lin, Xia, & Wan, 2010). Landfill management plays an important role on the critical issues of solid waste management, including global warming mitigation, biodiversity preservation, and land reclamation.

In recent years, information available to the mass public about both the causes and consequences of global warming and climate change has increased (Kellstedt, Zahran, & Vedlitz, 2008). Science educators have a significant role in empowering students to take action to reduce global warming (Skamp, Boyes, & Stanisstreet, 2013). Quantifying net global warming potential (NGWP) and greenhouse gas intensity

(GHGI) of an agricultural activity is a method to assess the mitigation potential of the activity (Pratibha et al., 2016). There are many opportunities to monitor the changes in freshwater fungi communities along latitudes and habitat gradients, and to study both ecological thresholds and consequences of such changes, particularly its feedback on the nutrient and carbon cycles in freshwater systems (Hyde, Fryar, Tian, Bahkali, & Xu, 2016).

Seo (2013) recommended the smart adaptation strategies (i.e., the environmental strategies used to reduce the damage from global warming and to mitigate greenhouse gases at the same time), such as natural resource uses, land use changes, consumer actions, migration, population changes, alternative energy sources, and technological advances. Significant reduction of CO_2 emissions from vehicles relates to clean technology innovations (de Stefano, Montes-Sancho, & Busch, 2016). Alternative fuels and vehicle technology can mitigate climate change and greenhouse gases from road transport (Nanaki & Koroneos, 2016).

There is a wealth of model-based evidence on the technology choices, costs, and other factors (e.g., fossil fuel demand) associated with mitigation toward stringent climate targets (Dessens, Anandarajah, & Gambhir, 2016). The simplified mathematical model has been developed to understand the effect of CO_2 on the mechanism of global warming, and the heat transfer in the atmosphere, particularly the radiation, can be described by the known relations in thermal engineering (Specht, Redemann, & Lorenz, 2016). Climate warming during the course of the 21st century is projected to be between 1.0 and 3.7 degree C depending on future greenhouse gas emissions, based on the ensemble-mean results of state-of-the-art Earth System Models (ESMs) (Anderson, Hawkins, & Jones, 2016).

Significant Aspects of Climate Change

Climate change has significant impacts on the native, threatened, and endangered wildlife (Htun, Gray, Lepczyk, Titmus, & Adams, 2016). Assessments of climate change impacts on species are needed for expecting potential biodiversity losses (Hellmann, Alkemade, & Knol, 2016). Ecological niche-based models are broadly utilized to map the habitat suitability of current and future potential distribution of species, using the precise coordination of species occurrences, along with climatic and various environmental variables (Chakraborty, Joshi, & Sachdeva, 2016).

Due to climate change, the shifts in germination phenology affect population dynamics, species composition, and community diversity (Walck, Hidayati, Dixon, Thompson, & Poschlod, 2015). Similar to climate change, species show much variation in their responses to the change of land use (Oliver & Morecroft, 2014). Climate change in combination with habitat destruction, degradation, and fragmentation may lead to the new waves of species extinctions in the near future as the species are unable to reach the cooler refuges due to altered landscapes (Cochard, 2011). To successfully manage for climate change, a better understanding will be needed of which species and systems will likely be most affected by climate change, how to preserve the evolutionary capacity of species, and how to implement adaptive management in modern systems (Lawler, 2009).

Building energy consumption is vulnerable to climate change due to the direct relationship between outside temperature and space cooling/heating (Huang & Gurney, 2016). Energy consumption and thermal comfort in buildings are heavily affected by weather conditions regarding climate change (Invidiata & Ghisi, 2016). The effect of climate change on the energy demand for office buildings is optimized by implementing the calculation procedure of ISO-13790:2008, based on the iterations of its envelope and pattern (Rubio-Bellido, Pérez-Fargallo, & Pulido-Arcas, 2016).

Policy options include implementation of BMPs through education and incentives for scale-dependent and site specific bioretention units to increase the resilience of the watershed system to the current and future climate change (Dudula & Randhir, 2016). The government policy intervention on affordability is essential for those encountering financial barriers during climate change (Lang, Radke, Chen, & Chan, 2016) and primary care physicians and public health officials must educate the public about the health relevance of climate change (Maibach et al., 2015).

Biophysical impacts of climate change extend the current climate trends, with high regional variability (Hopkins, Campbell-Hunt, Carter, Higham, & Rosin, 2015). Visualization aspects that are important for spurring reflection on adaptive action are specifying various climate parameters, relating climate impacts to the established practices for managing weather risks, and emphasizing diverse measures (Glaas, Ballantyne, Neset, & Linnér, 2017). The emissions of greenhouse gases are mainly from high-income countries, whereas the negative consequences of climate change are largely in low-income countries (Popescu & Nica, 2017). The logical chain of cause and effect from social drivers to CO_2 emission and climate change is used as an educational basis for advocating the global necessity and potential technological feasibility of CO_2 reduction (Ahamer, 2016).

Vulnerability assessments of climate change identify and characterize who and what are sensitive to climatic risks and why, characterize adaptive capacity and its determinants, and identify opportunities for adaptation (Ford et al., 2010) that can affect the species vulnerability toward climate change (Nadeau & Fuller, 2016). The spatial pattern of vulnerability identifies areas, requiring attention to adaptation action, whereas vulnerability assessment enables policy intervention and prioritization at local spatial scales (Kumar, Geneletti, & Nagendra, 2016). Determining where biodiversity is likely to be vulnerable to climate change and establishing the methods to reduce that vulnerability are the necessary steps to incorporate climate change into biodiversity management plans (Nadeau, Fuller, & Rosenblatt, 2015).

Climate change is expected to influence infection risks while bathing downstream of sewage emissions from combined sewage overflows (CSOs) or waste water treatment plants (WWTPs) due to the changes in pathogen influx, rising temperatures and changing the flow rates of the receiving waters (Sterk, de Man, Schijven, de Nijs, & de Roda Husman, 2016). Climate change obstructs the supply of water resources (Kumar, 2017) and the operation of rainwater harvesting system (Haque, Rahman, & Samali, 2016). Identifying and protecting the areas of the ocean may provide the important tool for the adaptation of climate change (Ban, Alidina, Okey, Gregg, & Ban, 2016). Due to climate change, sea level rise poses the great threats to all aquatic plant community types in the vicinity of the oceans, and the changes in weather patterns and salinity will affect many species (Short, Kosten, Morgan, Malone, & Moore, 2016).

Organisms alter their gene expression and metabolism to increase the concentrations of several antistress compounds and to change their physiology, phenology, growth, and reproduction in response to climate change (Peñuelas et al., 2013). Although tropical forests account for only a fraction of the planet's terrestrial surface, they exchange more CO_2 with the atmosphere than any other biome on the Earth, and thus play a disproportionate role in the global climate (Cavaleri, Reed, Smith, & Wood, 2015). The phase of the forest has a considerable impact on the effects of climate change on forest floor vegetation and its feedback effects on carbon cycles and on climate (Hedwall, Skoglund, & Linder, 2015).

Climate Change and Crop Production

The growing concern of the impact of climate change in forestry has prompted tree improvement programs and regulatory agencies to integrate climate change adaptation in the production and use of tree seed (Gray, Rweyongeza, Hamann, John, & Thomas, 2016). Deciduousness (i.e., the quality of trees which lose their leaves in winter or the dry season) in tropical trees has the potential to emerge as an important focus for ecological research to address the critical questions in global warming modeling, monitoring, and climate change (Singh & Kushwaha, 2016).

The impact of irrigation of bioenergy crops is the most prominent factor, leading to the higher water requirements under climate change mitigation if bioenergy crops are irrigated (Mouratiadou et al., 2016). Selection of prior crop with reduced vulnerability, increased climate change contribution rate, and reduced agricultural technology contribution rate can effectively address the adverse impacts of climate change and maximize climatic resources at a lower cost (Dong et al., 2016).

Climate Change and Risk Issues

Anthropogenic activities are responsible for the emission of gaseous and particulate pollutants that modify atmospheric composition (Maione et al., 2016). General beliefs (e.g., political orientations) and climate change-specific beliefs (e.g., believing in the anthropogenic causes of climate change) are the most prominent determinants of climate change-related risk perception (Safi, Smith, & Liu, 2012). Resilience management (e.g., risk retention) can be used to respond to the uncertainties in supply for adjusting production and mitigating the risks of climate change (Ho, Chen, Nobuyuki, Lur, & Lu, 2016).

Species range and climate change risk are often assessed through utilizing the species distribution models (SDM) that model the niche of species from presence points and environmental variables and project it in space and time (Fourcade, 2016). People's capacity to adapt to the shifting climate conditions is one of the most important characteristics to consider when addressing climate risks (Lohmann, 2016). The regulation of climate change must prioritize the risks of business-as-usual over the risks of change, must target the systemic change instead of stability, and must integrate those of individualization and compartmentalization (Heyvaert, 2011).

Public Perceptions of Climate Change

A substantial body of social science research examines the patterns of climate change views in the general public of countries around the world (McCright, Marquart-Pyatt, Shwom, Brechin, & Allen, 2016). Public engagement with climate change issues can facilitate energy efficiency policy but to succeed, wider climate policy measures are needed (Aune et al., 2016). The public understandings and perceptions of climate change have enhanced the interest of research and policy for almost three decades (Wolf & Moser, 2011). Citizens display a heavy reliance on the media as the sources of information, which are dominated by a techno-managerial discourse focusing on climate change (Carvalho, Schmidt, Santos, & Delicado, 2014). The changes in the public perceptions of climate change and trends in opinion can be consequential for national and international responses to climate change and can be underpinned by a range of social forces and physical phenomena (Capstick, Whitmarsh, Poortinga, Pidgeon, & Upham, 2015).

FUTURE RESEARCH DIRECTIONS

The classification of the extensive literature in the domains of global warming and climate change will provide the potential opportunities for future research. An examination of linkages among global warming, climate change, sustainability, sustainable consumption, risk assessment, and contingency planning in environmental settings should be further studied.

CONCLUSION

This chapter provided the advanced issues of global warming; the significant aspects of climate change; climate change and crop production; climate change and risk issues; and the public perceptions of climate change. Nowadays, global warming are climate change are the important environmental concerns that have a major impact on various environmental perspectives. The effects and impacts of global warming are climate change, sea level change, water balance, and human health. The problems of global warming and climate change can be controlled by minimizing the emission of greenhouse gases into the environment through conducting laws, reducing thermal power generating stations, planting trees, and sharing vehicles. Global warming and climate change must be recognized by local governments and environment-related international organizations in order to find the effective approaches to reducing their impacts in environmental settings.

REFERENCES

Abid, M., Schneider, U. A., & Scheffran, J. (2016). Adaptation to climate change and its impacts on food productivity and crop income: Perspectives of farmers in rural Pakistan. *Journal of Rural Studies*, *47*, 254–266. doi:10.1016/j.jrurstud.2016.08.005

Adhikari, U., Nejadhashemi, A. P., & Woznicki, S. A. (2015). Climate change and eastern Africa: A review of impact on major crops. *Food and Energy Security*, *4*(2), 110–132. doi:10.1002/fes3.61

Ahamer, G. (2016). Can educational approaches help to revolutionize quantitative solutions for climate change? In M. Raisinghani (Ed.), *Revolutionizing education through web-based instruction* (pp. 1–19). Hershey, PA: IGI Global. doi:10.4018/978-1-4666-9932-8.ch001

Anderson, T. R., Hawkins, E., & Jones, P. D. (2016). CO_2, the greenhouse effect and global warming: From the pioneering work of Arrhenius and Callendar to today's Earth System Models. *Endeavour*, *40*(3), 178–187. doi:10.1016/j.endeavour.2016.07.002 PMID:27469427

Arora, A., Rao, S. A., Chattopadhyay, R., Goswami, T., George, G., & Sabeerali, C. T. (2016). Role of Indian Ocean SST variability on the recent global warming hiatus. *Global and Planetary Change*, *143*, 21–30. doi:10.1016/j.gloplacha.2016.05.009

Aune, M., Godbolt, Å. L., Sørensen, K. H., Ryghaug, M., Karlstrøm, H., & Næss, R. (2016). Concerned consumption. Global warming changing household domestication of energy. *Energy Policy*, *98*, 290–297. doi:10.1016/j.enpol.2016.09.001

Ban, S. S., Alidina, H. M., Okey, T. A., Gregg, R. M., & Ban, N. C. (2016). Identifying potential marine climate change refugia: A case study in Canada's Pacific marine ecosystems. *Global Ecology and Conservation*, *8*, 41–54. doi:10.1016/j.gecco.2016.07.004

Bekele, I., & Ganpat, W. G. (2017). Education, extension, and training for climate change. In *Natural resources management: Concepts, methodologies, tools, and applications* (pp. 279–300). Hershey, PA: IGI Global. doi:10.4018/978-1-5225-0803-8.ch015

Borick, C. P., & Rabe, B. G. (2010). A reason to believe: Examining the factors that determine individual views on global warming. *Social Science Quarterly*, *91*(3), 777–800. doi:10.1111/j.1540-6237.2010.00719.x

Boussalis, C., Coan, T. G., & Poberezhskaya, M. (2016). Measuring and modeling Russian newspaper coverage of climate change. *Global Environmental Change*, *41*, 99–110. doi:10.1016/j.gloenvcha.2016.09.004

Capstick, S., Whitmarsh, L., Poortinga, W., Pidgeon, N., & Upham, P. (2015). International trends in public perceptions of climate change over the past quarter century. *Wiley Interdisciplinary Reviews: Climate Change*, *6*(1), 35–61. doi:10.1002/wcc.321

Carvalho, A., Schmidt, L., Santos, F. D., & Delicado, A. (2014). Climate change research and policy in Portugal. *Wiley Interdisciplinary Reviews: Climate Change*, *5*(2), 199–217. doi:10.1002/wcc.258

Cavaleri, M. A., Reed, S. C., Smith, W. K., & Wood, T. E. (2015). Urgent need for warming experiments in tropical forests. *Global Change Biology*, *21*(6), 2111–2121. doi:10.1111/gcb.12860 PMID:25641092

Chakraborty, A., Joshi, P. K., & Sachdeva, K. (2016). Predicting distribution of major forest tree species to potential impacts of climate change in the central Himalayan region. *Ecological Engineering*, *97*, 593–609. doi:10.1016/j.ecoleng.2016.10.006

Chaves, I. A., Melchers, R. E., Peng, L., & Stewart, M. G. (2016). Probabilistic remaining life estimation for deteriorating steel marine infrastructure under global warming and nutrient pollution. *Ocean Engineering*, *126*, 129–137. doi:10.1016/j.oceaneng.2016.09.013

Cochard, R. (2011). Consequences of deforestation and climate change on biodiversity. In Y. Trisurat, R. Shrestha, & R. Alkemade (Eds.), *Land use, climate change and biodiversity modeling: Perspectives and applications* (pp. 24–51). Hershey, PA: IGI Global. doi:10.4018/978-1-60960-619-0.ch002

Cortés-Borda, D., Ruiz-Hernández, A., Guillén-Gosálbez, G., Llop, M., Guimerà, R., & Sales-Pardo, M. (2015). Identifying strategies for mitigating the global warming impact of the EU-25 economy using a multi-objective input–output approach. *Energy Policy*, *77*, 21–30. doi:10.1016/j.enpol.2014.11.020

Danos, A., & Boulouta, K. (2011). The impact of climate change in the modern enterprise. *International Journal of Knowledge Society Research*, *2*(3), 26–35. doi:10.4018/jksr.2011070103 de Richter, R. K., Ming, T., Caillol, S., & Liu, W. (2016). Fighting global warming by GHG removal: Destroying CFCs and HCFCs in solar-wind power plant hybrids producing renewable energy with no-intermittency. *International Journal of Greenhouse Gas Control, 49*, 449–472. doi:10.1016/j.ijggc.2016.02.027

de Stefano, M. C., Montes-Sancho, M. J., & Busch, T. (2016). A natural resource-based view of climate change: Innovation challenges in the automobile industry. *Journal of Cleaner Production, 139*, 1436–1448. doi:10.1016/j.jclepro.2016.08.023

Dessens, O., Anandarajah, G., & Gambhir, A. (2016). Limiting global warming to 2 °C: What do the latest mitigation studies tell us about costs, technologies and other impacts? *Energy Strategy Reviews, 13-14*, 67–76. doi:10.1016/j.esr.2016.08.004

Dong, Z., Pan, Z., Wang, S., An, P., Zhang, J., Zhang, J., & Pan, X. et al. (2016). Effective crop structure adjustment under climate change. *Ecological Indicators, 69*, 571–577. doi:10.1016/j.ecolind.2016.04.010

dos Santos, M. A. O. (2012). Investigating consumer knowledge of global warming based on Rogers' knowledge stage of the innovation decision process. *International Journal of Consumer Studies, 36*(4), 385–393. doi:10.1111/j.1470-6431.2011.01069.x

Dudula, J., & Randhir, T. O. (2016). Modeling the influence of climate change on watershed systems: Adaptation through targeted practices. *Journal of Hydrology (Amsterdam), 541*, 703–713. doi:10.1016/j.jhydrol.2016.07.020

Elijido-Ten, E. O. (2017). Does recognition of climate change related risks and opportunities determine sustainability performance? *Journal of Cleaner Production, 141*, 956–966. doi:10.1016/j.jclepro.2016.09.136

Finke, T., Gilchrist, A., & Mouzas, S. (2016). Why companies fail to respond to climate change: Collective inaction as an outcome of barriers to interaction. *Industrial Marketing Management, 58*, 94–101. doi:10.1016/j.indmarman.2016.05.018

Ford, J. D., Keskitalo, E. C. H., Smith, T., Pearce, T., Berrang-Ford, L., Duerden, F., & Smit, B. (2010). Case study and analogue methodologies in climate change vulnerability research. *Wiley Interdisciplinary Reviews: Climate Change, 1*(3), 374–392. doi:10.1002/wcc.48

Forino, G., MacKee, J., & von Meding, J. (2016). A proposed assessment index for climate change-related risk for cultural heritage protection in Newcastle (Australia). *International Journal of Disaster Risk Reduction, 19*, 235–248. doi:10.1016/j.ijdrr.2016.09.003

Fourcade, Y. (2016). Comparing species distributions modelled from occurrence data and from expert-based range maps. Implication for predicting range shifts with climate change. *Ecological Informatics*, *36*, 8–14. doi:10.1016/j.ecoinf.2016.09.002

Glaas, E., Ballantyne, A. G., Neset, T. S., & Linnér, B. O. (2017). Visualization for supporting individual climate change adaptation planning: Assessment of a web-based tool. *Landscape and Urban Planning*, *158*, 1–11. doi:10.1016/j.landurbplan.2016.09.018

Goklany, I. M. (2012). Is climate change the number one threat to humanity? *Wiley Interdisciplinary Reviews: Climate Change*, *3*(6), 489–508. doi:10.1002/wcc.194

Goyal, M. K., & Royal, I. (2015). Soil carbon sequestration: An alternative option for climate change mitigation. In N. Gaurina-Medjimurec (Ed.), *Handbook of research on advancements in environmental engineering* (pp. 30–54). Hershey, PA: IGI Global. doi:10.4018/978-1-4666-7336-6.ch002

Graham, C. T., & Harrod, C. (2009). Implications of climate change for the fishes of the British Isles. *Journal of Fish Biology*, *74*(6), 1143–1205. doi:10.1111/j.1095-8649.2009.02180.x PMID:20735625

Gray, L. K., Rweyongeza, D., Hamann, A., John, S., & Thomas, B. R. (2016). Developing management strategies for tree improvement programs under climate change: Insights gained from long-term field trials with lodgepole pine. *Forest Ecology and Management*, *377*, 128–138. doi:10.1016/j.foreco.2016.06.041

Grillakis, M. G., Koutroulis, A. G., Seiradakis, K. D., & Tsanis, I. K. (2016). Implications of 2 °C global warming in European summer tourism. *Climate Services*, *1*, 30–38. doi:10.1016/j.cliser.2016.01.002

Guoju, X., Qiang, Z., Fengju, Z., Fei, M., Jing, W., Juying, H., & Zhengji, Q. et al. (2016). Warming influences the yield and water use efficiency of winter wheat in the semiarid regions of Northwest China. *Field Crops Research*, *199*, 129–135. doi:10.1016/j.fcr.2016.09.023

Hansen, J., Ruedy, R., Sato, M., & Lo, K. (2010). Global surface temperature change. *Reviews of Geophysics*, *48*(4), RG4004. doi:10.1029/2010RG000345

Haque, M. M., Rahman, A., & Samali, B. (2016). Evaluation of climate change impacts on rainwater harvesting. *Journal of Cleaner Production*, *137*, 60–69. doi:10.1016/j.jclepro.2016.07.038

Hedwall, P. O., Skoglund, J., & Linder, S. (2015). Interactions with successional stage and nutrient status determines the life-form-specific effects of increased soil temperature on boreal forest floor vegetation. *Ecology and Evolution*, *5*(4), 948–960. doi:10.1002/ece3.1412 PMID:25750720

Heino, J., Virkkala, R., & Toivonen, H. (2009). Climate change and freshwater biodiversity: Detected patterns, future trends and adaptations in northern regions. *Biological Reviews of the Cambridge Philosophical Society*, *84*(1), 39–54. doi:10.1111/j.1469-185X.2008.00060.x PMID:19032595

Hellmann, F., Alkemade, R., & Knol, O. M. (2016). Dispersal based climate change sensitivity scores for European species. *Ecological Indicators*, *71*, 41–46. doi:10.1016/j.ecolind.2016.06.013

Heyvaert, V. (2011). Governing climate change: Towards a new paradigm for risk regulation. *The Modern Law Review*, *74*(6), 817–844. doi:10.1111/j.1468-2230.2011.00874.x

Ho, C. H., Chen, J. L., Nobuyuki, Y., Lur, H. S., & Lu, H. J. (2016). Mitigating uncertainty and enhancing resilience to climate change in the fisheries sector in Taiwan: Policy implications for food security. *Ocean and Coastal Management*, *130*, 355–372. doi:10.1016/j.ocecoaman.2016.06.020

Hopkins, D., Campbell-Hunt, C., Carter, L., Higham, J. E. S., & Rosin, C. (2015). Climate change and Aotearoa New Zealand. *Wiley Interdisciplinary Reviews: Climate Change*, *6*(6), 559–583. doi:10.1002/wcc.355

Htun, H., Gray, S. A., Lepczyk, C. A., Titmus, A., & Adams, K. (2016). Combining watershed models and knowledge-based models to predict local-scale impacts of climate change on endangered wildlife. *Environmental Modelling & Software*, *84*, 440–457. doi:10.1016/j.envsoft.2016.07.009

Huang, J., & Gurney, K. R. (2016). The variation of climate change impact on building energy consumption to building type and spatiotemporal scale. *Energy*, *111*, 137–153. doi:10.1016/j.energy.2016.05.118

Hyde, K. D., Fryar, S., Tian, Q., Bahkali, A. H., & Xu, J. (2016). Lignicolous freshwater fungi along a north–south latitudinal gradient in the Asian/Australian region; can we predict the impact of global warming on biodiversity and function? *Fungal Ecology*, *19*, 190–200. doi:10.1016/j.funeco.2015.07.002

Invidiata, A., & Ghisi, E. (2016). Impact of climate change on heating and cooling energy demand in houses in Brazil. *Energy and Building*, *130*, 20–32. doi:10.1016/j.enbuild.2016.07.067

Kaarlejärvi, E., Hoset, K. S., & Olofsson, J. (2015). Mammalian herbivores confer resilience of Arctic shrub-dominated ecosystems to changing climate. *Global Change Biology*, *21*(9), 3379–3388. doi:10.1111/gcb.12970 PMID:25967156

Kellstedt, P. M., Zahran, S., & Vedlitz, A. (2008). Personal efficacy, the information environment, and attitudes toward global warming and climate change in the United States. *Risk Analysis*, *28*(1), 113–126. doi:10.1111/j.1539-6924.2008.01010.x PMID:18304110

Kumar, C. P. (2017). Impact of climate change on groundwater resources. In *Natural resources management: Concepts, methodologies, tools, and applications* (pp. 1094–1120). Hershey, PA: IGI Global. doi:10.4018/978-1-5225-0803-8.ch052

Kumar, P., Geneletti, D., & Nagendra, H. (2016). Spatial assessment of climate change vulnerability at city scale: A study in Bangalore, India. *Land Use Policy*, *58*, 514–532. doi:10.1016/j.landusepol.2016.08.018

Lang, W., Radke, J. D., Chen, T., & Chan, E. H. W. (2016). Will affordability policy transcend climate change? A new lens to re-examine equitable access to healthcare in the San Francisco Bay Area. *Cities (London, England)*, *58*, 124–136. doi:10.1016/j.cities.2016.05.014

Lawler, J. J. (2009). Climate change adaptation strategies for resource management and conservation planning. *Annals of the New York Academy of Sciences*, *1162*(1), 79–98. doi:10.1111/j.1749-6632.2009.04147.x PMID:19432646

Li, G., Han, H., Du, Y., Hui, D., Xia, J., Niu, S., & Wan, S. et al. (2017). Effects of warming and increased precipitation on net ecosystem productivity: A long-term manipulative experiment in a semiarid grassland. *Agricultural and Forest Meteorology*, *232*, 359–366. doi:10.1016/j.agrformet.2016.09.004

Li, H., & Wang, X. (2012). Study on Chinese low carbon economic model. In D. Ura & P. Ordóñez de Pablos (Eds.), *Advancing technologies for Asian business and economics: Information management developments* (pp. 220–228). Hershey, PA: IGI Global. doi:10.4018/978-1-4666-0276-2.ch017

Li, J., Luo, Y., Natali, S., Schuur, E. A. G., Xia, J., Kowalczyk, E., & Wang, Y. (2014). Modeling permafrost thaw and ecosystem carbon cycle under annual and seasonal warming at an Arctic tundra site in Alaska. *Journal of Geophysical Research. Biogeosciences*, *119*(6), 1129–1146. doi:10.1002/2013JG002569

Lin, D., Xia, J., & Wan, S. (2010). Climate warming and biomass accumulation of terrestrial plants: A meta-analysis. *The New Phytologist*, *188*(1), 187–198. doi:10.1111/j.1469-8137.2010.03347.x PMID:20609113

Liu, Y., Mu, J., Niklas, K. J., Li, G., & Sun, S. (2012). Global warming reduces plant reproductive output for temperate multi-inflorescence species on the Tibetan plateau. *The New Phytologist*, *195*(2), 427–436. doi:10.1111/j.1469-8137.2012.04178.x PMID:22591333

Lohmann, H. (2016). Comparing vulnerability and adaptive capacity to climate change in individuals of coastal Dominican Republic. *Ocean and Coastal Management*, *132*, 111–119. doi:10.1016/j.ocecoaman.2016.08.009

López-Moreno, J. I., Boike, J., Sanchez-Lorenzo, A., & Pomeroy, J. W. (2016). Impact of climate warming on snow processes in Ny-Ålesund, a polar maritime site at Svalbard. *Global and Planetary Change*, *146*, 10–21. doi:10.1016/j.gloplacha.2016.09.006

Maibach, E. W., Kreslake, J. M., Roser-Renouf, C., Rosenthal, S., Feinberg, G., & Leiserowitz, A. A. (2015). Do Americans understand that global warming is harmful to human health? Evidence from a national survey. *Annals of Global Health*, *81*(3), 396–409. doi:10.1016/j.aogh.2015.08.010 PMID:26615074

Maione, M., Fowler, D., Monks, P. S., Reis, S., Rudich, Y., Williams, M. L., & Fuzzi, S. (2016). Air quality and climate change: Designing new win-win policies for Europe. *Environmental Science & Policy*, *65*, 48–57. doi:10.1016/j.envsci.2016.03.011

McCright, A. M., Marquart-Pyatt, S. T., Shwom, R. L., Brechin, S. R., & Allen, S. (2016). Ideology, capitalism, and climate: Explaining public views about climate change in the United States. *Energy Research & Social Science*, *21*, 180–189. doi:10.1016/j.erss.2016.08.003

Mouratiadou, I., Biewald, A., Pehl, M., Bonsch, M., Baumstark, L., Klein, D., & Kriegler, E. et al. (2016). The impact of climate change mitigation on water demand for energy and food: An integrated analysis based on the shared socioeconomic pathways. *Environmental Science & Policy*, *64*, 48–58. doi:10.1016/j.envsci.2016.06.007

Mukherjee, S., & Mukerjee, A. (2017). Sustainable business development by responding to climate change: A case of the Tata group. In *Natural resources management: Concepts, methodologies, tools, and applications* (pp. 416–431). Hershey, PA: IGI Global. doi:10.4018/978-1-5225-0803-8.ch021

Mullan, D., Vandaele, K., Boardman, J., Meneely, J., & Crossley, L. H. (2016). Modelling the effectiveness of grass buffer strips in managing muddy floods under a changing climate. *Geomorphology*, *270*, 102–120. doi:10.1016/j.geomorph.2016.07.012

Nadeau, C. P., & Fuller, A. K. (2016). Combining landscape variables and species traits can improve the utility of climate change vulnerability assessments. *Biological Conservation*, *202*, 30–38. doi:10.1016/j.biocon.2016.07.030

Nadeau, C. P., Fuller, A. K., & Rosenblatt, D. L. (2015). Climate-smart management of biodiversity. *Ecosphere*, *6*(6), 1–17. doi:10.1890/ES15-00069.1

Nanaki, E. A., & Koroneos, C. J. (2016). Climate change mitigation and deployment of electric vehicles in urban areas. *Renewable Energy*, *99*, 1153–1160. doi:10.1016/j.renene.2016.08.006

Oliver, T. H., & Morecroft, M. D. (2014). Interactions between climate change and land use change on biodiversity: Attribution problems, risks, and opportunities. *Wiley Interdisciplinary Reviews: Climate Change*, *5*(3), 317–335. doi:10.1002/wcc.271

Pappis, C. P. (2011). Global warming: Basic facts. In C. Pappis (Ed.), *Climate change, supply chain management and enterprise adaptation: Implications of global warming on the economy* (pp. 30–65). Hershey, PA: IGI Global. doi:10.4018/978-1-61692-800-1.ch002

Pearce, W., Brown, B., Nerlich, B., & Koteyko, N. (2015). Communicating climate change: Conduits, content, and consensus. *Wiley Interdisciplinary Reviews: Climate Change*, *6*(6), 613–626. doi:10.1002/wcc.366

Peñuelas, J., Sardans, J., Estiarte, M., Ogaya, R., Carnicer, J., Coll, M., & Jump, A. S. et al. (2013). Evidence of current impact of climate change on life: A walk from genes to the biosphere. *Global Change Biology*, *19*(8), 2303–2338. doi:10.1111/gcb.12143 PMID:23505157

Perreault, P., & Shur, Y. (2016). Seasonal thermal insulation to mitigate climate change impacts on foundations in permafrost regions. *Cold Regions Science and Technology, 132*, 7–18. doi:10.1016/j.coldregions.2016.09.008

Ploberger, C. (2013). A critical assessment of environmental degeneration and climate change: A multidimensional (political, economic, social) challenge for China's future economic development. In Z. Luo (Ed.), *Technological solutions for modern logistics and supply chain management* (pp. 212–229). Hershey, PA: IGI Global. doi:10.4018/978-1-4666-2773-4.ch014

Popescu, G. H., & Nica, E. (2017). Global warming, climate policy, and the green paradox. In *Natural resources management: Concepts, methodologies, tools, and applications* (pp. 1–19). Hershey, PA: IGI Global. doi:10.4018/978-1-5225-0803-8.ch001

Popović, V., & Mijajlović, N. (2014). Climate change and sustainable development in agriculture and forestry. In *Sustainable practices: Concepts, methodologies, tools, and applications* (pp. 18–48). Hershey, PA: IGI Global. doi:10.4018/978-1-4666-4852-4.ch002

Pratibha, G., Srinivas, I., Rao, K. V., Shanker, A. K., Raju, B. M. K., Choudhary, D. K., & Maheswari, M. et al. (2016). Net global warming potential and greenhouse gas intensity of conventional and conservation agriculture system in rainfed semi arid tropics of India. *Atmospheric Environment, 145*, 239–250. doi:10.1016/j.atmosenv.2016.09.039

Pries, C. E. H., van Logtestijn, R. S. P., Schuur, E. A. G., Natali, S. M., Cornelissen, J. H. C., Aerts, R., & Dorrepaal, E. (2015). Decadal warming causes a consistent and persistent shift from heterotrophic to autotrophic respiration in contrasting permafrost ecosystems. *Global Change Biology, 21*(12), 4508–4519. doi:10.1111/gcb.13032 PMID:26150277

Ranney, M. A., & Clark, D. (2016). Climate change conceptual change: Scientific information can transform attitudes. *Topics in Cognitive Science, 8*(1), 49–75. doi:10.1111/tops.12187 PMID:26804198

Reynolds, T. W., Bostrom, A., Read, D., & Morgan, M. G. (2010). Now what do people know about global climate change? Survey studies of educated laypeople. *Risk Analysis, 30*(10), 1520–1538. doi:10.1111/j.1539-6924.2010.01448.x PMID:20649942

Rubio-Bellido, C., Pérez-Fargallo, A., & Pulido-Arcas, J. A. (2016). Optimization of annual energy demand in office buildings under the influence of climate change in Chile. *Energy*, *114*, 569–585. doi:10.1016/j.energy.2016.08.021

Safi, A. S., Smith, W. J. Jr, & Liu, Z. (2012). Rural Nevada and climate change: Vulnerability, beliefs, and risk perception. *Risk Analysis*, *32*(6), 1041–1059. doi:10.1111/j.1539-6924.2012.01836.x PMID:22583075

Santamouris, M., Cartalis, C., Synnefa, A., & Kolokotsa, D. (2015). On the impact of urban heat island and global warming on the power demand and electricity consumption of buildings: A review. *Energy and Building*, *98*, 119–124. doi:10.1016/j.enbuild.2014.09.052

Schaefer, K., Zhang, T., Bruhwiler, L., & Barrett, A. P. (2011). Amount and timing of permafrost carbon release in response to climate warming. *Tellus. Series B, Chemical and Physical Meteorology*, *63*(2), 165–180. doi:10.1111/j.1600-0889.2011.00527.x

Schliephack, J., & Dickinson, J. E. (2017). Tourists' representations of coastal managed realignment as a climate change adaptation strategy. *Tourism Management*, *59*, 182–192. doi:10.1016/j.tourman.2016.08.004

Schmeida, M., & McNeal, R. S. (2017). U.S. public support to climate change initiatives?: Setting stricter carbon dioxide emission limits on power plants. In *Natural resources management: Concepts, methodologies, tools, and applications* (pp. 1196–1215). Hershey, PA: IGI Global. doi:10.4018/978-1-5225-0803-8.ch058

Seo, S. N. (2013). Economics of global warming as a global public good: Private incentives and smart adaptations. *Regional Science Policy & Practice*, *5*(1), 83–95. doi:10.1111/j.1757-7802.2012.01088.x

Short, F. T., Kosten, S., Morgan, P. A., Malone, S., & Moore, G. E. (2016). Impacts of climate change on submerged and emergent wetland plants. *Aquatic Botany*, *135*, 3–17. doi:10.1016/j.aquabot.2016.06.006

Singh, K. P., & Kushwaha, C. P. (2016). Deciduousness in tropical trees and its potential as indicator of climate change: A review. *Ecological Indicators*, *69*, 699–706. doi:10.1016/j.ecolind.2016.04.011

Skamp, K., Boyes, E., & Stanisstreet, M. (2013). Beliefs and willingness to act about global warming: Where to focus science pedagogy? *Science Education*, *97*(2), 191–217. doi:10.1002/sce.21050

Specht, E., Redemann, T., & Lorenz, N. (2016). Simplified mathematical model for calculating global warming through anthropogenic CO_2. *International Journal of Thermal Sciences*, *102*, 1–8. doi:10.1016/j.ijthermalsci.2015.10.039

Srivastava, N. (2017). Climate change mitigation: Collective efforts and responsibly. In *Natural resources management: Concepts, methodologies, tools, and applications* (pp. 64–76). Hershey, PA: IGI Global. doi:10.4018/978-1-5225-0803-8.ch003

Sterk, A., de Man, H., Schijven, J. F., de Nijs, T., & de Roda Husman, A. M. (2016). Climate change impact on infection risks during bathing downstream of sewage emissions from CSOs or WWTPs. *Water Research*, *105*, 11–21. doi:10.1016/j.watres.2016.08.053 PMID:27591704

Walck, J. L., Hidayati, S. N., Dixon, K. W., Thompson, K., & Poschlod, P. (2015). Climate change and plant regeneration from seed. *Global Change Biology*, *17*(6), 2145–2161. doi:10.1111/j.1365-2486.2010.02368.x

Wang, C., & Dong, S. (2010). Is the basin-wide warming in the North Atlantic Ocean related to atmospheric carbon dioxide and global warming? *Geophysical Research Letters*, *37*(8), L08707. doi:10.1029/2010GL042743

Weng, Y. C., Fujiwara, T., Houng, H. J., Sun, C. H., Li, W. Y., & Kuo, Y. W. (2015). Management of landfill reclamation with regard to biodiversity preservation, global warming mitigation and landfill mining: Experiences from the Asia–Pacific region. *Journal of Cleaner Production*, *104*(1), 364–373. doi:10.1016/j.jclepro.2015.05.014

Winter, R. A. (2014). Innovation and the dynamics of global warming. *Journal of Environmental Economics and Management*, *68*(1), 124–140. doi:10.1016/j.jeem.2014.01.005

Wolf, J., & Moser, S. C. (2011). Individual understandings, perceptions, and engagement with climate change: Insights from in-depth studies across the world. *Wiley Interdisciplinary Reviews: Climate Change*, *2*(4), 547–569. doi:10.1002/wcc.120

Yumino, S., Uchida, T., Sasaki, K., Kobayashi, H., & Mochida, A. (2015). Total assessment for various environmentally conscious techniques from three perspectives: Mitigation of global warming, mitigation of UHIs, and adaptation to urban warming. *Sustainable Cities and Society*, *19*, 236–249. doi:10.1016/j.scs.2015.05.010

Zhang, M., Li, B., & Xiong, Z. Q. (2016). Effects of organic fertilizer on net global warming potential under an intensively managed vegetable field in southeastern China: A three-year field study. *Atmospheric Environment*, *145*, 92–103. doi:10.1016/j.atmosenv.2016.09.024

Zhao, X., Leiserowitz, A. A., Maibach, E. W., & Roser-Renouf, C. (2011). Attention to science/environment news positively predicts and attention to political news negatively predicts global warming risk perceptions and policy support. *Journal of Communication*, *61*(4), 713–731. doi:10.1111/j.1460-2466.2011.01563.x

ADDITIONAL READING

Atienza, D., Sabatés, A., Isari, S., Saiz, E., & Calbet, A. (2016). Environmental boundaries of marine Cladoceran distributions in the NW Mediterranean: Implications for their expansion under global warming. *Journal of Marine Systems*, *164*, 30–41. doi:10.1016/j.jmarsys.2016.08.003

Bai, E., Li, S., Xu, W., Li, W., Dai, W., & Jiang, P. (2013). A meta-analysis of experimental warming effects on terrestrial nitrogen pools and dynamics. *The New Phytologist*, *199*(2), 441–451. doi:10.1111/nph.12252 PMID:23550663

Cai, X., Zhang, X., Noël, P. H., & Shafiee-Jood, M. (2015). Impacts of climate change on agricultural water management: A review. *Wiley Interdisciplinary Reviews: Water*, *2*(5), 439–455. doi:10.1002/wat2.1089

Carey, M. (2012). Climate and history: A critical review of historical climatology and climate change historiography. *Wiley Interdisciplinary Reviews: Climate Change*, *3*(3), 233–249. doi:10.1002/wcc.171

Carlsson, M., Naroznova, I., Møller, J., Scheutz, C., & Lagerkvist, A. (2015). Importance of food waste pre-treatment efficiency for global warming potential in life cycle assessment of anaerobic digestion systems. *Resources, Conservation and Recycling*, *102*, 58–66. doi:10.1016/j.resconrec.2015.06.012

Corner, A., Roberts, O., Chiari, S., Völler, S., Mayrhuber, E. S., Mandl, S., & Monson, K. (2015). How do young people engage with climate change? The role of knowledge, values, message framing, and trusted communicators. *Wiley Interdisciplinary Reviews: Climate Change*, *6*(5), 523–534. doi:10.1002/wcc.353

Ekström, M., Grose, M. R., & Whetton, P. H. (2015). An appraisal of downscaling methods used in climate change research. *Wiley Interdisciplinary Reviews: Climate Change*, *6*(3), 301–319. doi:10.1002/wcc.339

Haluza-DeLay, R. (2014). Religion and climate change: Varieties in viewpoints and practices. *Wiley Interdisciplinary Reviews: Climate Change*, *5*(2), 261–279. doi:10.1002/wcc.268

Hegerl, G., & Zwiers, F. (2011). Use of models in detection and attribution of climate change. *Wiley Interdisciplinary Reviews: Climate Change*, *2*(4), 570–591. doi:10.1002/wcc.121

Herman, B. C. (2015). The influence of global warming science views and sociocultural factors on willingness to mitigate global warming. *Science Education*, *99*(1), 1–38. doi:10.1002/sce.21136

Hufnagel, E. (2015). Preservice elementary teachers' emotional connections and disconnections to climate change in a science course. *Journal of Research in Science Teaching*, *52*(9), 1296–1324. doi:10.1002/tea.21245

Launius, R. D. (2011). Climate change and spaceflight: An historiographical review. *Wiley Interdisciplinary Reviews: Climate Change*, *2*(3), 412–427. doi:10.1002/wcc.114

Leichenko, R., & Silva, J. A. (2014). Climate change and poverty: Vulnerability, impacts, and alleviation strategies. *Wiley Interdisciplinary Reviews: Climate Change*, *5*(4), 539–556. doi:10.1002/wcc.287

Lewis, K. H., & Lenton, T. M. (2015). Knowledge problems in climate change and security research. *Wiley Interdisciplinary Reviews: Climate Change*, *6*(4), 383–399. doi:10.1002/wcc.346

Liu, Y., Zhou, Z., Zhang, X., Xu, X., Chen, H., & Xiong, Z. (2015). Net global warming potential and greenhouse gas intensity from the double rice system with integrated soil–crop system management: A three-year field study. *Atmospheric Environment*, *116*(1), 92–101. doi:10.1016/j.atmosenv.2015.06.018

McNeall, D., Halloran, P. R., Good, P., & Betts, R. A. (2011). Analyzing abrupt and nonlinear climate changes and their impacts. *Wiley Interdisciplinary Reviews: Climate Change*, *2*(5), 663–686. doi:10.1002/wcc.130

Moser, S. C. (2014). Communicating adaptation to climate change: The art and science of public engagement when climate change comes home. *Wiley Interdisciplinary Reviews: Climate Change*, *5*(3), 337–358. doi:10.1002/wcc.276

Nadeau, C. P., & Fuller, A. K. (2015). Accounting for multiple climate components when estimating climate change exposure and velocity. *Methods in Ecology and Evolution*, *6*(6), 697–705. doi:10.1111/2041-210X.12360

Peel, J., & Osofsky, H. M. (2013). Climate change litigation's regulatory pathways: A comparative analysis of the United States and Australia. *Law & Policy*, *35*(3), 150–183. doi:10.1111/lapo.12003

Rickards, L., Wiseman, J., & Kashima, Y. (2014). Barriers to effective climate change mitigation: The case of senior government and business decision makers. *Wiley Interdisciplinary Reviews: Climate Change*, *5*(6), 753–773. doi:10.1002/wcc.305

Rosenzweig, C., & Neofotis, P. (2013). Detection and attribution of anthropogenic climate change impacts. *Wiley Interdisciplinary Reviews: Climate Change*, *4*(2), 121–150. doi:10.1002/wcc.209

Samson, J., Berteaux, D., McGill, B. J., & Humphries, M. M. (2011). Geographic disparities and moral hazards in the predicted impacts of climate change on human populations. *Global Ecology and Biogeography*, *20*(4), 532–544. doi:10.1111/j.1466-8238.2010.00632.x

Schaefli, B. (2015). Projecting hydropower production under future climates: A guide for decision-makers and modelers to interpret and design climate change impact assessments. *Wiley Interdisciplinary Reviews: Water*, *2*(4), 271–289. doi:10.1002/wat2.1083

Schurer, A. P., Hegerl, G. C., & Obrochta, S. P. (2015). Determining the likelihood of pauses and surges in global warming. *Geophysical Research Letters*, *42*(14), 5974–5982. doi:10.1002/2015GL064458

Scott, D., Gössling, S., & Hall, C. M. (2012). International tourism and climate change. *Wiley Interdisciplinary Reviews: Climate Change*, *3*(3), 213–232. doi:10.1002/wcc.165

Shuman, B. (2012). Patterns, processes, and impacts of abrupt climate change in a warm world: The past 11,700 years. *Wiley Interdisciplinary Reviews: Climate Change*, *3*(1), 19–43. doi:10.1002/wcc.152

Smith, N., & Leiserowitz, A. (2014). The role of emotion in global warming policy support and opposition. *Risk Analysis*, *34*(5), 937–948. doi:10.1111/risa.12140 PMID:24219420

Soam, S., Kapoor, M., Kumar, R., Borjesson, P., Gupta, R. P., & Tuli, D. K. (2016). Global warming potential and energy analysis of second generation ethanol production from rice straw in India. *Applied Energy*, *184*, 353–364. doi:10.1016/j.apenergy.2016.10.034

Spence, A., Poortinga, W., & Pidgeon, N. (2012). The psychological distance of climate change. *Risk Analysis*, *32*(6), 957–972. doi:10.1111/j.1539-6924.2011.01695.x PMID:21992607

Vallis, G. K., Zurita-Gotor, P., Cairns, C., & Kidston, J. (2015). Response of the large-scale structure of the atmosphere to global warming. *Quarterly Journal of the Royal Meteorological Society*, *141*(690), 1479–1501. doi:10.1002/qj.2456

Weng, Z., Haque, N., Mudd, G. M., & Jowitt, S. M. (2016). Assessing the energy requirements and global warming potential of the production of rare earth elements. *Journal of Cleaner Production*, *139*, 1282–1297. doi:10.1016/j.jclepro.2016.08.132

Wood, T. E., Cavaleri, M. A., & Reed, S. C. (2012). Tropical forest carbon balance in a warmer world: A critical review spanning microbial- to ecosystem-scale processes. *Biological Reviews of the Cambridge Philosophical Society*, *87*(4), 912–927. doi:10.1111/j.1469-185X.2012.00232.x PMID:22607308

KEY TERMS AND DEFINITIONS

Carbon Dioxide: A colorless and odorless gas that is produced through combustion and respiration.

Climate: The weather of a location over time or the environment.

Climate Change: The change in the Earth's climate that is produced by global warming.

Ecosystem: A system made up of a community of animals, plants, and bacteria interrelated together with its physical and chemical environment.

Emission: The discharge of some substances, such as a gas, or the thing that is discharged.

Global Warming: The continuing slight increase in the Earth's average temperature which scientists believe will result in climate change throughout the world.

Greenhouse Gas: Any gas that contributes to the greenhouse effect when released into the atmosphere.

Mitigation: The process of making something less severe or less intense.

Species: A group of animals, plants, or other living things that share the common characteristics.

Chapter 5
Pollution and Renewable Energy:
Advanced Issues and Aspects

Kijpokin Kasemsap
Suan Sunandha Rajabhat University, Thailand

ABSTRACT

This chapter presents the overview of pollution; the issues of soil pollution, water pollution, and air pollution; the aspects of renewable energy; energy security and energy imports; and renewable energy policy and renewable energy policy instruments. Pollution is one of the most important environmental, social, and health issues in the world. Pollution creates many diseases and causes death of many people across the globe. The environmental damage caused by pollution can reach catastrophic proportions and destroy entire ecosystems leading to the death of many species and a big biodiversity loss. Renewable energy is a critical part of reducing global carbon emissions and the pace of investment has greatly increased as the cost of technologies fall and efficiency continues to rise. Renewable energy offers a wide variety of different options to choose from as countries can choose between sun, wind, biomass, geothermal energy, and water resources.

INTRODUCTION

The industrialization of society and modernization of society's way of living have resulted in the introduction of a large number of man-made substances in the environment, many pollutants can be harmful to both human health and other species (Voulvoulis & Georges, 2016). Industrial pollution is one of the largest

DOI: 10.4018/978-1-5225-3379-5.ch005

environmental and livelihood issues faced by the developing societies (Charuvilayil, 2013). Achieving a sustainable urban environment requires accounting for the economic, environmental, and social impacts of the development to reduce the pollutions of air, soil, water, odor, and noise (Hammad, Akbarnezhad, & Rey, 2016) toward obtaining a green environment (Fagbenro, 2016).

Entry of toxic heavy metals and minerals in human systems mainly through contaminated water, food, and air, leads to health problems (Rathoure, 2016). Toxic effects of heavy metals are dependent on the concentration of metals, reactivity of metal species, and duration of exposure (Sharma, Kaur, Katnoria, & Nagpal, 2016). Companies using hazardous materials in their production have started to consider the environmentally integrated manufacturing systems to effectively decrease their impacts on environment and to prevent the environmental pollution at source (Kasap, Demirer, & Ergün, 2013) through ecotechnology focusing on minimizing the consequences of pollution by manipulating ecosystem-internal processes (Benndorf, 2008).

Global warming problem caused by greenhouse gas emissions (Kuang et al., 2016) from burning fossil fuels has caused wide public concern all over the world (Zhang et al., 2015). In recent years, renewable energy (e.g., wind energy, geothermal energy, solar energy, biomass energy, and hydropower) has increasingly become the world's strategic choice to solve environmental pollution, to address the energy crisis, and to achieve social sustainable development (He, Xu, Pang, Tian, & Wu, 2016). Depletion of fossil fuel resources, fluctuation in the crude oil prices, and emersion of new environmental problems due to the greenhouse gasses effects of fossil fuel combustion have convinced many governments to invest in the development of power generation based on renewable and sustainable energy (RSE) resources (Hosseini & Abdul Wahid, 2014).

Local governments around the world pursue a wide range of activities to reduce local greenhouse gas emissions, to increase energy efficiency, and to promote the locally generated renewable energy (Radzi, 2015). Renewable energy development is influenced by regulatory institutions, the party affiliations of both governor and legislators, and the professionalism of the legislature, accompanied by the effects of various policy instruments (Yi & Feiock, 2014). Government support and commitment are of particular importance for renewable energy technology innovation activities, which are highly contingent on policy and market uncertainty (Liang & Fiorino, 2013). The development, transfer, and use of renewable energy technologies are the promising ways toward sustainable development (Pueyo & Linares, 2012).

This chapter is based on a literature review of pollution and renewable energy. The extensive literature of pollution and renewable energy provides a contribution to practitioners and researchers by presenting the advanced issues and aspects of pollution and renewable energy.

Background

Recent trade and environmental policy debates seem to take as given that regulatory stringency in developed countries shifts polluting industries to the developing world (Levinson & Taylor, 2008). The exponential growth of urban settings has led to an increase in pollutants and waste management issues (Hua, 2016) toward ecological destruction and environmental pollution (Li & Zhang, 2012). Cities with higher levels of manufacturing agglomeration are associated with more serious environmental degradation, whereas cities with higher levels of service agglomeration show a tendency of higher environmental quality (Sun & Yuan, 2015).

Environmental management is a systematic strategy that companies can apply to find the different ways for saving water, energy, and materials, and for reducing the negative environmental impacts (Kasemsap, 2017a). Environmental policies should be adopted by regions or nations to regulate the effective environment toward achieving the quality of living (Chao, Laffargue, Liu, Sgro, & Xiao, 2015). The less-developed regions should take advantage of economic growth to invest more advanced environment protection technologies (Li, Han, Li, Lu, & Zhao, 2016). Many developing countries' national plans and urbanization policies lack the effective measures to address the environmental degradation and enhance the sustainable utilization of natural resources (Küçükali, 2016).

Traditional energy sources or fossil fuel resources (e.g., oil, coal, and natural gas) are the largest energy sources for human living and production (Li, He, & Li, 2016) and are not statistically significant in renewable energy deployment (Kilinc-Ata, 2016). Lobbying activities of traditional energy sources are restraining the deployment of renewable energy (Marques & Fuinhas, 2011). Renewable promotion policies are being enacted due to the powerful lobbying activities in traditional industries (Aguirre & Ibikunle, 2014). Financial assistance and technology transfer supported by governments (Li, Rubin, & Onyina, 2013) are required to promote their efforts toward a climate-friendly sustainable economy (Urban, 2009).

Higher income countries are relatively capable of sustaining the costs of renewable energy technologies and stimulate renewable energy deployment through economic incentives (Aguirre & Ibikunle, 2014). Concerning renewable energy projects, smart-grid policies include a new generation of regulations and finance models, such as regulatory targets, requirements for data security, renewable energy credits, and various interconnection tariffs (Brown & Zhou, 2013). In grid-connected renewable energy, many investors face difficulty in evaluating proper return, making them more averse to financing renewable energy projects, affecting transborder project development opportunities (Fay & Kumar, 2013). If the technical progress and development speed of renewable energy lags behind the growth in demand, it will be difficult to realize the improvement of its energy structure (Shi, 2009).

ADVANCES ISSUES AND ASPECTS OF POLLUTION AND RENEWABLE ENERGY

This section provides the overview of pollution; the issues of soil pollution, water pollution, and air pollution; the aspects of renewable energy; energy security and energy imports; and renewable energy policy and renewable energy policy instruments.

Overview of Pollution

In an era of closer worldwide economic integration, the role that environmental regulations play in shaping a country's comparative advantage is greater than ever (Elliott & Shimamoto, 2008). The implementation of both economic policy instrument (e.g., pollution discharge fee) and public participation (e.g., letter complaints on environmental problems) encourages the industrial relocation, whereas the implementation of environmental legal policy instrument (e.g., environmental laws, regulations, and rules) prevents the polluting industries from relocating to other regions (Zheng & Shi, 2017).

Whenever the production economy grows, the stock of pollution increases (Gutierrez, 2008). The effective control of environmental pollution requires the strategic changes in tax structure (Zhang & Zheng, 2009) and the development of individual's human capital (Guangming & Zhaofeng, 2013). The focus of research on pollution control has frequently been in response to regulatory drivers, examining the efficacy, efficiency, and economic incentives of particular approaches (Wilson, 2015). The environmental risk assessment is based on calculating the probability for an ecosystem to receive a dose of pollutant (Hrncevic, 2015). There are different shapes of the relationship between income growth and various measures of pollution and environmental degradation, and developing countries must recognize that no one size fits all in this relationship (Ayadi, 2010).

Issues of Soil Pollution

Soil contamination is one the major soil threats, especially in regions with a high population density and strong industrialization (Biasioli, Fabietti, Barberis, & Ajmone-Marsan, 2012). Contaminated soil is the soil in which human or natural activity has increased the content of harmful substances whose concentrations are harmful to soil health (Gómez-Sagasti, Epelde, Alkorta, & Garbisu, 2016) and human activity for the production of plants (Kisić, 2015).

Soil contaminants can be classified as macro and micro contaminants by quantity, while micro-contaminants divide into two groups: organic and inorganic pollutants (Sheikhavandi, 2015). Pesticides are an important class of soil contaminants because of their intentional toxicity and widespread applications to home, garden, and agricultural soils. Pesticides are difficult to regulate because they are marketed in thousands of products made from hundreds of potentially toxic

The choice of an appropriate sampling scheme is a crucial step in the process of soil pollution assessment and risk management (Rocco et al., 2016). Correctly distinguishing between natural and anthropogenic trace metal contents in soils is crucial for evaluating soil contamination (Desaules, 2012). Suitable remediation techniques (e.g., chemical stabilization of soils by amendments) can be used for reducing the bioavailability of soil contaminants (Ettler, 2016).

Issues of Water Pollution

Water pollution occurs when pollutants are directly or indirectly discharged into water bodies without water treatment process to remove the harmful compounds (Jaiswal & Chattopadhyaya, 2015) toward the illnesses of community members (McKeown, 2017). Construction site-related water pollution causes the irreversible damages to the surrounding environment with additional cost, time, and resources required for rectification works (Belayutham, González, & Yiu, 2016).

The quality of surface waters and groundwater of a geographical region can be affected by anthropogenic activities, land use patterns, and fecal pollution sources from humans and animals (Tran, Gin, & Ngo, 2015). To guarantee the security of water quality in water transfer channels, especially in open channels, analysis of potential emergent pollution sources in the water transfer process is critical (Tang, Yi, Yang, & Sun, 2016).

The reduced water pollution can increase health outcome (Wang & Yang, 2016) and food safety (Lu et al., 2015). Thus, there is a need for unconventional methods to provide the better tools for the assessment and management of water quality problems to adopt management policies and set the limits for sustainable drainage water reuse toward reducing water pollution (Sallam, Youssef, Embaby, & Shaltot, 2011). Sediments are an important repository for various pollutants (e.g., pesticides and heavy metals) and play a significant role as the sensitive indicators for monitoring contaminants in aquatic systems (Bat & Özkan, 2015).

Issues of Air Pollution

One of the impacts of rapid economic growth is on the quality of air, which deteriorates with the release of emissions from different sources (Sharma, 2016). Air pollution is contamination of the indoor or outdoor environment by any chemical, physical or biological agent that modifies the natural characteristics of the atmosphere (Bogdanović & Lazarević, 2017). Sources of air pollution include natural phenomena (e.g., volcanoes, wildfires, and land dust) and human activities (Franchini, Guida, Tufano, & Coppola, 2012). Perception of air pollution by the general public (Saksena, 2011) is an important issue in the development of risk assessment studies and policies (Deguen, Ségala, Pédrono, & Mesbah, 2012).

Acute effects of air pollution regarding the exposures to nitrogen dioxide, nitrogen oxide, sulfur dioxide, and fine particulate matter (Chen, Wan, Yang, & Zou, 2015) are reflected in respiratory symptoms, cardiovascular events, hospitalizations and mortality, whereas long-term effects include the reduced spirometric lung function in adults, cardiovascular disease, and lung cancer (Berend, 2016). With the harmful effects of air pollution on health, public health measures are urgently needed to improve the air quality in order to reduce the morbidity and mortality of patients with the disabling disease (Ko & Hui, 2012). A greater understanding of the adverse health consequences of exposures to air pollution in early life (Tong, 2016) is required to encourage policymakers to reduce such exposures (Goldizen, Sly, & Knibbs, 2016) and to reduce health risks (Johnson, 2012).

The availability of precise information enables businesses to take preventive measures against the emission of poisonous gases and other hazards, thus improving the safety of personnel and equipment (Roy & Bandyopadhyay, 2011). Data mining techniques can help investigate the behavior of ambient air pollutants and allow practitioners to extract the useful knowledge from the complex air quality data (Kim, Temiyasathit, Park, & Chen, 2009). Making mathematic models to simulate the environment changes (e.g., diffusion of gas in the air) is very helpful for environmental impact assessment (EIA) and environmental protection (Meng, Fahong, & Lei, 2008).

Aspects of Renewable Energy

Renewable energy is categorized as the alternative energy in the opposite to the fossil fuels (Dresselhaus & Tomas, 2001). Climate change involves the changes in the world's weather, in particular the fact that it is believed to be getting warmer as a result of human activity increasing the level of carbon dioxide in the atmosphere (Kasemsap, 2017b). An awareness of an impending crisis and concern over climate change are driving an increase in research and development (R&D) for alternative energy sources instead of fossil-based energy (Jang, Lee, & Oh, 2013). The development of alternative energy is affected by the acquired technology and the weather conditions (Mathieson, Henrik Hund, & Karlson, 2011). Alternative energy resources must be developed to prevent the adverse effects on the environment and the economy (Wang, Sung, & Hsu, 2016).

Traditional energy sources carry significant environmental problems, such as climate change, air pollution, and habitat destruction (Kilinc-Ata, 2016). Renewable energy proves to be a strategic investment, more friendly to people and environment than traditional energy which causes health problems, such as casual respiratory diseases and cancer (McKenzie, Witter, Newman, & Adgate, 2012) and many environmental problems facing humanity, such as greenhouse gas emissions, air pollution, water pollution, and soil contamination. The main cause of carbon dioxide released into the atmosphere are the fossil fuels (Archer et al., 2009) used to produce both electricity and heat, which leads to global warming.

Energy use is recognized as an impact category within life cycle assessment (Arvidsson & Svanström, 2016). The mitigation of climate change demands the decarbonization of the energy supply of industrialized countries by 2050 (Wiese, Bökenkamp, Wingenbach, & Hohmeyer, 2014). In response to the threat of climate change, many governments have set the policy goals to rapidly increase the utilization of renewable energy in order to lessen the reliance upon fossil fuels and reduce the emissions of greenhouse gases (Devine-Wright, 2011). Renewable energy alleviates the depletion of traditional energy sources (Li et al., 2016) and meets the requirement for enhancing economic development (Pacesila, Burcea, & Colesca, 2016).

Government should promote renewable energy development through supporting technology research, thus encouraging information dissemination, introducing effective administrative policies, and accomplishing incentive mechanisms (Uyterlinde, Junginger, de Vries, Faaij, & Turkenburg, 2007). Tradeoffs among renewable energy management (REM) system costs, renewable energy availabilities, and energy-shortage risks can be tackled with the consideration of climate change, which would have both positive and negative impacts on the REM system cost, energy supply, and greenhouse gas emission (Cai, Huang, Yeh, Liu, & Li, 2012).

Renewable energy exploitation contributes to addressing environmental problems, preserving natural resources, and improving the quality of life (Stigka, Paravantis, & Mihalakakou, 2014). Public acceptability often poses a barrier toward renewable energy development (Devine-Wright, 2005). The energy needs of contemporary society can be satisfied through the utilization of renewable energy, because it limits the environmental impact of conventional electricity generators and reduces the emission of pollutants (Fagiani, Barquín, & Hakvoort, 2013). To promote renewable energy consumption, economic welfare of the population should be enhanced (Ackah, Alabi, & Lartey, 2016).

Renewable energy is eco-friendly and sustainable (Wang et al., 2016). District heating system plays an important role in the implementation of future sustainable energy systems (Lund et al., 2014). The 4th generation district heating (4GDH) is an integrated part of the operation of smart energy systems, such as the integrated smart electricity, gas, and thermal grids (Lund, Möller, Mathiesen, & Dyrelund, 2010). Smart energy systems enable renewable energy (e.g., wind, solar, wave power, and low-level heat sources) have a flexibility in energy storage (Li et al., 2016).

Energy Security and Energy Imports

Energy security has an impact on renewable energy deployment (Dong, 2012). Security of energy supply is an essential ingredient for any nation's survival, and the insufficient energy sources contribute to the increased vulnerability of national economies to the fluctuations in the prices of energy products (Pacesila et al., 2016). It is crucial for a country with high energy imports to enhance energy security by increasing renewable energy deployment (Kilinc-Ata, 2016).

Renewable energy can reduce the society's dependence on fossil fuels and geopolitical risks concerning import dependency (del Rio & Burguillo, 2009). Energy imports are related to the reduced renewable energy utilization after controlling for the imported energy and electricity (Popp, Hascic, & Medhi, 2011). Most countries invest in renewable energy not only to reduce dependence on the imported oil, but also to minimize the price volatility related to the imports of fossil fuel toward increasing energy security (Menyah & Wolde-Rufael, 2010).

Renewable Energy Policy and Renewable Energy Policy Instruments

Energy situation varies from country to country and from region to region (Pacesila et al., 2016). Many governments cope with energy challenges, such as demand growth, national security risk with fossil fuel dependence, climate change, and pollution (Stokes, 2013). The performance of the policy depends on the policy design and the environment of the policy, such as macroeconomic conditions, regulatory risk, and financial sources (Recalde, 2016). Both national innovation policy and an international treaty establishing a low-emission technology commitment should be the central focus of renewable energy-related climate change policy (Karlsson & Symons, 2015).

The deeper reforms in electricity generation are consistent with more ambitious renewable energy policy (Kim, Yang, & Urpelainen, 2016). In smart grid regarding electricity generation, the integration of renewable energy sources (e.g., solar energy and wind energy) is a challenging task because of their intermittent nature (Rasheed et al., 2016). The European Union (EU), the largest economy in the world, has the sustainable energy goals for the year 2020 (Lund, 2012) and has established a target of 20% of electricity generation from renewable energy sources by 2020 (Menegaki, 2013) in terms of the proportion of electricity generation, primary energy consumption (i.e., the total energy consumed by a country), and final energy consumption (Lean & Smyth, 2013) toward the reduction of greenhouse gas emissions (Sikkema et al., 2011). Furthermore, in the US states, Oregon's target is 25% of electricity from renewable energy sources by 2025 (Delmas & Montes-Sancho, 2011).

Strategic renewable energy policies promote the diffusion of renewable energy sources within the market (Kilinc-Ata, 2016) and encourage both investors and producers to respond to market developments without increasing the costs for consumers (Zamfir, Colesca, & Corbos, 2016). Increasing the availability and reducing the cost of renewable sources require well-coordinated government policy intervention (Jones, 2009). Quotas are quantity-based policy instruments, and they usually require electricity retailers to supply a minimum percentage of electricity demand from renewable energy sources (Buckman, 2011). Renewable portfolio standard (RPS) policies express the great political feasibility toward making new renewable technologies become cost-competitive with conventional sources of fossil fuel energy (Rabe, 2008).

Tax deduction is the most effective financial policy instrument to promote the consumer's acceptance of renewable energy (Sardianou & Genoudi, 2013). Kanes and Wohlgemuth (2008) indicated that a fossil energy tax reduction is more efficient and useful than subsidy and tax reduction for renewable energy, required to encourage the efficient investment decisions. Chang et al. (2009) suggested that there is a positive relationship between traditional energy prices and renewable energy growth.

FUTURE RESEARCH DIRECTIONS

The classification of the extensive literature in the domains of pollution and renewable energy will provide the potential opportunities for future research. Corporate social responsibility (CSR) is a company's sense of responsibility toward the community and environment in which it operates (Kasemsap, 2017c). Promoting sustainable consumption is equally important to limit negative environmental and social externalities as well as to provide markets for sustainable products (Kasemsap, 2017d). Lean supply chain and green supply chain strategies help companies maximize the improvement of lean production in operations management (Kasemsap, 2016a).

Concerning green supply chain management (GSCM) practices, the generation of hazardous substances can be reduced, thus preventing organizations as a result of violating environmental regulations (Kasemsap, 2016b). Sustainable supply chain management (SSCM) involves integrating the environmentally and financially viable practices into the complete supply chain lifecycle, from product design and development, to material selection, manufacturing, packaging, transportation, warehousing, distribution, consumption, return, and disposal (Kasemsap, 2017e). An examination of linkages among pollution, renewable energy, CSR, sustainable consumption, GSCM, and SSCM would seem to be viable for future research efforts.

CONCLUSION

This chapter explained the overview of pollution; the issues of soil pollution, water pollution, and air pollution; the aspects of renewable energy; energy security and energy imports; and renewable energy policy and renewable energy policy instruments. Pollution issue is one of the most important environmental, social, and health issues in the world. Pollution creates many diseases and causes death of many people across the globe. The environmental damage caused by pollution can reach catastrophic proportions and destroy entire ecosystems leading to the death of many species and a big biodiversity loss.

Industrial pollution contaminates many sources of drinking water, releases unwanted toxins into the air, and reduces the quality of soil all over the world. When involved in an array of processes, the water comes into contact with heavy metals, harmful chemicals, radioactive waste, and organic sludge. Air pollution occurs when any harmful gases, dust, smoke enters into the atmosphere and makes it difficult for plants, animals, and humans to survive as the air becomes dirty. Air pollution disturbs the natural balance of gases in the atmosphere, having serious ecological implications. Soil pollution creates many problems in agriculture and destroys local vegetation. Soil pollution causes the chronic health issues to the people that come in contact with such soil on a daily basis.

Renewable energy is a critical part of reducing global carbon emissions and the pace of investment has greatly increased as the cost of technologies fall and efficiency continues to rise. Renewable energy offers a wide variety of different options to choose from as countries can choose between sun, wind, biomass, geothermal energy, and water resources. Renewable energy facilities generally require less maintenance than traditional generators. Their fuel being derived from natural and available resources reduces the costs of operation. Renewable energy produces little or no waste products (e.g., carbon dioxide or other chemical pollutants) and has minimal impact on the environment. In addition, renewable energy projects can bring the economic benefits to many regional areas, as most projects are located away from large urban centers and suburbs of the capital cities.

REFERENCES

Ackah, I., Alabi, O., & Lartey, A. (2016). Analysing the efficiency of renewable energy consumption among oil-producing African countries. *OPEC Energy Review*, *40*(3), 316–334.

Aguirre, M., & Ibikunle, G. (2014). Determinants of renewable energy growth: A global sample analysis. *Energy Policy*, *69*, 374–384.

Archer, D., Eby, M., Brovkin, V., Ridgwell, A., Cao, L., & Mikolajewicz, U. et al.. (2009). Atmospheric lifetime of fossil fuel carbon dioxide. *Annual Review of Earth and Planetary Sciences*, *37*(1), 117–134.

Arvidsson, R., & Svanström, M. (2016). A framework for energy use indicators and their reporting in life cycle assessment. *Integrated Environmental Assessment and Management*, *12*(3), 429–436. PMID:26551582

Ayadi, F. S. (2010). An empirical investigation of environmental Kuznets curve in Nigeria. *International Journal of Green Computing*, *1*(2), 31–39.

Bat, L., & Özkan, E. Y. (2015). Heavy metal levels in sediment of the Turkish Black Sea coast. In I. Zlateva, V. Raykov, & N. Nikolov (Eds.), *Progressive engineering practices in marine resource management* (pp. 399–419). Hershey, PA: IGI Global.

Belayutham, S., González, V. A., & Yiu, T. W. (2016). The dynamics of proximal and distal factors in construction site water pollution. *Journal of Cleaner Production*, *113*, 54–65.

Benndorf, J. (2008). Ecotechnology and emission control: Alternative or mutually promoting strategies in water resources management? *International Review of Hydrobiology*, *93*(4/5), 466–478.

Berend, N. (2016). Contribution of air pollution to COPD and small airway dysfunction. *Respirology (Carlton, Vic.)*, *21*(2), 237–244. PMID:26412571

Biasioli, M., Fabietti, G., Barberis, R., & Ajmone-Marsan, F. (2012). An appraisal of soil diffuse contamination in an industrial district in northern Italy. *Chemosphere*, *88*(10), 1241–1249. PMID:22608707

Bogdanović, D., & Lazarević, K. (2017). Early warning system and adaptation advice to reduce human health consequences of extreme weather conditions and air pollution. In *Public health and welfare: Concepts, methodologies, tools, and applications* (pp. 527–564). Hershey, PA: IGI Global.

Brown, M. A., & Zhou, S. (2013). Smart-grid policies: An international review. *Wiley Interdisciplinary Reviews: Energy and Environment*, 2(2), 121–139.

Buckman, G. (2011). The effectiveness of renewable portfolio standard banding and carve-outs in supporting high-cost types of renewable electricity. *Energy Policy*, *39*(7), 4105–4114.

Cai, Y. P., Huang, G. H., Yeh, S. C., Liu, L., & Li, G. C. (2012). A modeling approach for investigating climate change impacts on renewable energy utilization. *International Journal of Energy Research*, *36*(6), 764–777.

Cetin, M., Seker, F., & Cavlak, H. (2015). The impact of trade openness on environmental pollution: A panel cointegration and causality analysis. In E. Sorhun, Ü. Hacıoğlu, & H. Dinçer (Eds.), *Regional economic integration and the global financial system* (pp. 221–232). Hershey, PA: IGI Global.

Chang, T. H., Huang, C. M., & Lee, M. C. (2009). Threshold effect of the economic growth rate on the renewable energy development from a change in energy price: Evidence from OECD countries. *Energy Policy*, *37*(12), 5796–5802.

Chao, C. C., Laffargue, J. P., Liu, X., Sgro, P. M., & Xiao, Y. (2015). Migration and the environment: Policy reform in a polluted open economy. *World Economy*, *38*(1), 48–62.

Charuvilayil, R. A. (2013). Industrial pollution and people's movement: A case study of Eloor Island Kerala, India. In H. Muga & K. Thomas (Eds.), *Cases on the diffusion and adoption of sustainable development practices* (pp. 312–351). Hershey, PA: IGI Global.

Chen, G., Wan, X., Yang, G., & Zou, X. (2015). Traffic-related air pollution and lung cancer: A meta-analysis. *Thoracic Cancer*, *6*(3), 307–318. PMID:26273377

Deguen, S., Ségala, C., Pédrono, G., & Mesbah, M. (2012). A new air quality perception scale for global assessment of air pollution health effects. *Risk Analysis*, *32*(12), 2043–2054. PMID:22852801

del Rio, P., & Burguillo, M. (2009). An empirical analysis of the impact of renewable energy deployment on local sustainability. *Renewable & Sustainable Energy Reviews*, *13*(6/7), 1314–1325.

Delmas, M., & Montes-Sancho, M. (2011). U.S. state policies for renewable energy: Context and effectiveness. *Energy Policy*, *39*(5), 2273–2288.

Desaules, A. (2012). Critical evaluation of soil contamination assessment methods for trace metals. *The Science of the Total Environment*, *426*, 120–131. PMID:22542230

Devine-Wright, P. (2005). Beyond NIMBYism: Towards an integrated framework for understanding public perceptions of wind energy. *Wind Energy (Chichester, England)*, *8*(2), 125–139.

Devine-Wright, P. (2011). Public engagement with large-scale renewable energy technologies: Breaking the cycle of NIMBYism. *Wiley Interdisciplinary Reviews: Climate Change*, *2*(1), 19–26.

Dong, C. G. (2012). Feed-in tariff vs. renewable portfolio standard: An empirical test of their relative effectiveness in promoting wind capacity development. *Energy Policy*, *42*, 476–485.

Dresselhaus, M. S., & Tomas, I. L. (2001). Alternative energy technologies. *Nature*, *414*(6861), 332–337. PMID:11713539

Elliott, R. J. R., & Shimamoto, K. (2008). Are ASEAN countries havens for Japanese pollution-intensive industry? *World Economy*, *31*(2), 236–254.

Ettler, V. (2016). Soil contamination near non-ferrous metal smelters: A review. *Applied Geochemistry*, *64*, 56–74.

Fagbenro, O. K. (2016). Leachate pollution and impact to environment. In H. Aziz & S. Amr (Eds.), *Control and treatment of landfill leachate for sanitary waste disposal* (pp. 173–199). Hershey, PA: IGI Global.

Fagiani, R., Barquín, J., & Hakvoort, R. (2013). Risk-based assessment of the cost-efficiency and the effectivity of renewable energy support schemes: Certificate markets versus feed-in tariffs. *Energy Policy*, *55*, 648–661.

Fay, J., & Kumar, U. (2013). An index-based model for determining the investment benchmark of renewable energy projects in South Africa. *The South African Journal of Economics*, *81*(3), 416–426.

Franchini, M., Guida, A., Tufano, A., & Coppola, A. (2012). Air pollution, vascular disease and thrombosis: Linking clinical data and pathogenic mechanisms. *Journal of Thrombosis and Haemostasis*, *10*(12), 2438–2451. PMID:23006215

Goldizen, F. C., Sly, P. D., & Knibbs, L. D. (2016). Respiratory effects of air pollution on children. *Pediatric Pulmonology*, *51*(1), 94–108. PMID:26207724

Gómez-Sagasti, M. T., Epelde, L., Alkorta, I., & Garbisu, C. (2016). Reflections on soil contamination research from a biologist's point of view. *Applied Soil Ecology*, *105*, 207–210.

Guangming, L., & Zhaofeng, A. (2013). Empirical study on the correlations of environmental pollution, human capital, and economic growth: Based on the 1990-2007 data in Guangdong China. In P. Ordóñez de Pablos (Ed.), *Green technologies and business practices: An IT approach* (pp. 128–137). Hershey, PA: IGI Global.

Gutierrez, M. J. (2008). Dynamic inefficiency in an overlapping generation economy with pollution and health costs. *Journal of Public Economic Theory*, *10*(4), 563–594.

Hammad, A., Akbarnezhad, A., & Rey, D. (2016). Accounting for noise pollution in planning of smart cities. In G. Hua (Ed.), *Smart cities as a solution for reducing urban waste and pollution* (pp. 149–196). Hershey, PA: IGI Global.

He, Y., Xu, Y., Pang, Y., Tian, H., & Wu, R. (2016). A regulatory policy to promote renewable energy consumption in China: Review and future evolutionary path. *Renewable Energy*, *89*, 695–705.

Hosseini, S. E., & Abdul Wahid, M. (2014). The role of renewable and sustainable energy in the energy mix of Malaysia: A review. *International Journal of Energy Research*, *38*(14), 1769–1792.

Hrncevic, L. (2015). Petroleum industry environmental performance and risk. In I. Management Association (Ed.), Transportation systems and engineering: Concepts, methodologies, tools, and applications (pp. 32–56). Hershey, PA: IGI Global.

Hua, G. B. (2016). *Smart cities as a solution for reducing urban waste and pollution* (pp. 1–362). Hershey, PA: IGI Global.

Jaiswal, A., & Chattopadhyaya, M. C. (2015). Water pollution and its treatment. In N. Gaurina-Medjimurec (Ed.), *Handbook of research on advancements in environmental engineering* (pp. 147–171). Hershey, PA: IGI Global.

Jang, Y. S., Lee, D. J., & Oh, H. S. (2013). Evaluation of new and renewable energy technologies in Korea using real options. *International Journal of Energy Research*, *37*(13), 1645–1656.

Jennings, A. A., & Li, Z. (2014). Scope of the worldwide effort to regulate pesticide contamination in surface soils. *Journal of Environmental Management*, *146*, 420–443. PMID:25199603

Jones, S. (2009). The future of renewable energy in Australia: A test for cooperative federalism? *Australian Journal of Public Administration*, *68*(1), 1–20.

Kanes, D. A., & Wohlgemuth, N. (2008). Evaluation of renewable energy policies in an integrated economic energy environment model. *Forest Policy and Economics*, *10*(3), 128–139.

Karlsson, R., & Symons, J. (2015). Making climate leadership meaningful: Energy research as a key to global decarbonisation. *Global Policy*, *6*(2), 107–117.

Kasap, S., Demirer, S. U., & Ergün, S. (2013). An environmentally integrated manufacturing analysis combined with waste management in a car battery manufacturing plant. In I. Management Association (Ed.), Industrial engineering: Concepts, methodologies, tools, and applications (pp. 907–932). Hershey, PA: IGI Global.

Kasemsap, K. (2016a). The roles of lean and green supply chain management strategies in the global business environments. In S. Joshi & R. Joshi (Eds.), *Designing and implementing global supply chain management* (pp. 152–173). Hershey, PA: IGI Global.

Kasemsap, K. (2016b). Multifaceted applications of green supply chain management. In A. Jean-Vasile (Ed.), *Food science, production, and engineering in contemporary economies* (pp. 327–354). Hershey, PA: IGI Global.

Kasemsap, K. (2017a). Environmental management and waste management: Principles and applications. In U. Akkucuk (Ed.), *Ethics and sustainability in global supply chain management* (pp. 26–49). Hershey, PA: IGI Global.

Kasemsap, K. (2017b). Sustainability, environmental sustainability, and sustainable tourism: Advanced issues and implications. In N. Ray (Ed.), *Business infrastructure for sustainability in developing economies* (pp. 1–24). Hershey, PA: IGI Global.

Kasemsap, K. (2017c). Corporate social responsibility: Theory and applications. In A. Bhattacharya (Ed.), *Strategic human capital development and management in emerging economies* (pp. 188–218). Hershey, PA: IGI Global.

Kasemsap, K. (2017d). Mastering consumer attitude and sustainable consumption in the digital age. In N. Suki (Ed.), *Handbook of research on leveraging consumer psychology for effective customer engagement* (pp. 16–41). Hershey, PA: IGI Global.

Kasemsap, K. (2017e). Advocating sustainable supply chain management and sustainability in global supply chain. In M. Khan, M. Hussain, & M. Ajmal (Eds.), *Green supply chain management for sustainable business practice* (pp. 234–271). Hershey, PA: IGI Global.

Kilinc-Ata, N. (2016). The evaluation of renewable energy policies across EU countries and US states: An econometric approach. *Energy for Sustainable Development, 31*, 83–90.

Kim, S. B., Temiyasathit, C., Park, S., & Chen, V. C. (2009). Spatio-temporal data mining for air pollution problems. In J. Wang (Ed.), *Encyclopedia of data warehousing and mining* (2nd ed.; pp. 1815–1822). Hershey, PA: IGI Global.

Kim, S. E., Yang, J., & Urpelainen, J. (2016). Does power sector deregulation promote or discourage renewable energy policy? Evidence from the States, 1991–2012. *The Review of Policy Research, 33*(1), 22–50.

Kisić, I. (2015). Effects of soil contamination on the selection of remediation method. In N. Gaurina-Medjimurec (Ed.), *Handbook of research on advancements in environmental engineering* (pp. 200–227). Hershey, PA: IGI Global.

Ko, F. W. S., & Hui, D. S. C. (2012). Air pollution and chronic obstructive pulmonary disease. *Respirology (Carlton, Vic.), 17*(3), 395–401. PMID:22142380

Kuang, Y., Zhang, Y., Zhou, B., Li, C., Cao, Y., Li, L., & Zeng, L. (2016). A review of renewable energy utilization in islands. *Renewable & Sustainable Energy Reviews, 59*, 504–513.

Küçükali, U. F. (2016). Ecological influences on the evolving planning system in Turkey. In U. Benna & S. Garba (Eds.), *Population growth and rapid urbanization in the developing world* (pp. 298–312). Hershey, PA: IGI Global.

Lean, H. H., & Smyth, R. (2013). Will policies to promote renewable electricity generation be effective? Evidence from panel stationarity and unit root tests for 115 countries. *Renewable & Sustainable Energy Reviews, 22*, 371–379.

Levinson, A., & Taylor, M. S. (2008). Unmasking the pollution haven effect. *International Economic Review, 49*(1), 223–254.

Li, D., He, J., & Li, L. (2016). A review of renewable energy applications in buildings in the hot-summer and warm-winter region of China. *Renewable & Sustainable Energy Reviews, 57*, 327–336.

Li, H., & Zhang, X. (2012). Study on environmental tax: A case of China. In D. Ura & P. Ordóñez de Pablos (Eds.), *Advancing technologies for Asian business and economics: Information management developments* (pp. 207–219). Hershey, PA: IGI Global.

Li, T., Han, Y., Li, Y., Lu, Z., & Zhao, P. (2016). Urgency, development stage and coordination degree analysis to support differentiation management of water pollution emission control and economic development in the eastern coastal area of China. *Ecological Indicators*, *71*, 406–415.

Li, W., Rubin, T. H., & Onyina, P. A. (2013). Comparing solar water heater popularization policies in China, Israel and Australia: The roles of governments in adopting green innovations. *Sustainable Development*, *21*(3), 160–170.

Liang, J., & Fiorino, D. J. (2013). The implications of policy stability for renewable energy innovation in the United States, 1974–2009. *Policy Studies Journal: the Journal of the Policy Studies Organization*, *41*(1), 97–118.

Lu, Y., Song, S., Wang, R., Liu, Z., Meng, J., & Sweetman, A. J. et al.. (2015). Impacts of soil and water pollution on food safety and health risks in China. *Environment International*, *77*, 5–15. PMID:25603422

Lund, H., Möller, B., Mathiesen, B. V., & Dyrelund, A. (2010). The role of district heating in future renewable energy systems. *Energy*, *35*(3), 1381–1390.

Lund, H., Werner, S., Wiltshire, R., Svendsen, S., Thorsen, J. E., Hvelplund, F., & Mathiesen, B. V. (2014). 4th generation district heating (4GDH): Integrating smart thermal grids into future sustainable energy systems. *Energy*, *68*, 1–11.

Lund, P. (2012). The European Union challenge: Integration of energy, climate, and economic policy. *Wiley Interdisciplinary Reviews: Energy and Environment*, *1*(1), 60–68.

Marques, C. A., & Fuinhas, J. A. (2011). Do energy efficiency measures promote the use of renewable sources? *Environmental Science & Policy*, *14*(4), 471–481.

Mathieson, B. V., Henrik Hund, B. V., & Karlson, K. (2011). 100% renewable energy system, climate mitigation and economic growth. *Applied Energy*, *88*(2), 488–501.

McKenzie, L. M., Witter, R. Z., Newman, L. S., & Adgate, J. L. (2012). Human health risk assessment of air emissions from development of unconventional natural gas resources. *The Science of the Total Environment*, *424*(1), 79–87. PMID:22444058

McKeown, A. E. (2017). Nurses, healthcare, and environmental pollution and solutions: Breaking the cycle of harm. In *Public health and welfare: Concepts, methodologies, tools, and applications* (pp. 110–133). Hershey, PA: IGI Global.

Menegaki, A. N. (2013). Growth and renewable energy in Europe: Benchmarking with data envelopment analysis. *Renewable Energy*, *60*, 363–369.

Meng, Z., Fahong, Z., & Lei, L. (2008). Information technology and environment. In Y. Kurihara, S. Takaya, H. Harui, & H. Kamae (Eds.), *Information technology and economic development* (pp. 201–212). Hershey, PA: IGI Global.

Menyah, K., & Wolde-Rufael, Y. (2010). CO_2 emissions, nuclear energy, renewable energy and economic growth in the US. *Energy Policy*, *38*(6), 2911–2915.

Pacesila, M., Burcea, S. G., & Colesca, S. E. (2016). Analysis of renewable energies in European Union. *Renewable & Sustainable Energy Reviews*, *56*, 156–170.

Popp, D., Hascic, I., & Medhi, N. (2011). Technology and the diffusion of renewable energy. *Energy Economics*, *33*(4), 648–662.

Pueyo, A., & Linares, P. (2012). Renewable technology transfer to developing countries: One size does not fit all. *IDS Working Papers, 2012*(412), 1–39.

Rabe, B. G. (2008). States on steroids: The intergovernmental odyssey of American climate policy. *The Review of Policy Research*, *25*(2), 105–128.

Radzi, A. (2015). A survey of expert attitudes on understanding and governing energy autonomy at the local level. *Wiley Interdisciplinary Reviews: Energy and Environment*, *4*(5), 397–405.

Rasheed, M. B., Javaid, N., Ahmad, A., Awais, M., Khan, Z. A., Qasim, U., & Alrajeh, N. (2016). Priority and delay constrained demand side management in real-time price environment with renewable energy source. *International Journal of Energy Research*, *40*(14), 2002–2021.

Rathoure, A. K. (2016). Heavy metal pollution and its management: Bioremediation of heavy metal. In A. Rathoure & V. Dhatwalia (Eds.), *Toxicity and waste management using bioremediation* (pp. 27–50). Hershey, PA: IGI Global.

Recalde, M. Y. (2016). The different paths for renewable energies in Latin American countries: The relevance of the enabling frameworks and the design of instruments. *Wiley Interdisciplinary Reviews: Energy and Environment*, *5*(3), 305–326.

Rocco, C., Duro, I., di Rosa, S., Fagnano, M., Fiorentino, N., Vetromile, A., & Adamo, P. (2016). Composite vs. discrete soil sampling in assessing soil pollution of agricultural sites affected by solid waste disposal. *Journal of Geochemical Exploration*, *170*, 30–38.

Roy, S. D. A., & Bandyopadhyay, S. (2011). Testbed implementation of a pollution monitoring system using wireless sensor network for the protection of public spaces. In V. Sridhar & D. Saha (Eds.), *Recent advances in broadband integrated network operations and services management* (pp. 263–276). Hershey, PA: IGI Global.

Saksena, S. (2011). Public perceptions of urban air pollution risks. *Risk, Hazards & Crisis in Public Policy*, *2*(1), 1–19.

Sallam, G. A., Youssef, T., Embaby, M. E., & Shaltot, F. (2011). Using geographic information system to infollow the fertilizers pollution migration. In *Green technologies: Concepts, methodologies, tools and applications* (pp. 564–586). Hershey, PA: IGI Global.

Sardianou, E., & Genoudi, P. (2013). Which factors affect the willingness of consumers to adopt renewable energies? *Renewable Energy*, *57*, 1–4.

Sharma, A., Kaur, M., Katnoria, J. K., & Nagpal, A. K. (2016). Heavy metal pollution: A global pollutant of rising concern. In A. Rathoure & V. Dhatwalia (Eds.), *Toxicity and waste management using bioremediation* (pp. 1–26). Hershey, PA: IGI Global.

Sharma, S. (2016). Economic damage due to ozone pollution in NCR: Ozone impacts. In A. Goswami & A. Mishra (Eds.), *Economic modeling, analysis, and policy for sustainability* (pp. 284–306). Hershey, PA: IGI Global.

Sheikhavandi, T. (2015). Soil contamination. In N. Gaurina-Medjimurec (Ed.), *Handbook of research on advancements in environmental engineering* (pp. 173–199). Hershey, PA: IGI Global.

Shi, D. (2009). Analysis of China's renewable energy development under the current economic and technical circumstances. *China & World Economy*, *17*(2), 94–109.

Sikkema, R., Steiner, M., Junginger, M., Hiegl, W., Hansen, M. T., & Faaij, A. (2011). The European wood pellet markets: Current status and prospects for 2020. *Biofuels, Bioproducts & Biorefining*, *5*(3), 250–278.

Stigka, E. K., Paravantis, J. A., & Mihalakakou, G. K. (2014). Social acceptance of renewable energy sources: A review of contingent valuation applications. *Renewable & Sustainable Energy Reviews*, *32*, 100–106.

Stokes, C. L. (2013). The politics of renewable energy policies: The case of feed-in tariffs in Ontario, Canada. *Energy Policy*, *56*, 490–500.

Sun, P., & Yuan, Y. (2015). Industrial agglomeration and environmental degradation: Empirical evidence in Chinese cities. *Pacific Economic Review*, *20*(4), 544–568.

Tang, C., Yi, Y., Yang, Z., & Sun, J. (2016). Risk analysis of emergent water pollution accidents based on a Bayesian network. *Journal of Environmental Management*, *165*, 199–205. PMID:26433361

Tong, H. (2016). Dietary and pharmacological intervention to mitigate the cardiopulmonary effects of air pollution toxicity. *Biochimica et Biophysica Acta (BBA) -. General Subjects*, *1860*(12), 2891–2898. PMID:27189803

Tran, N. H., Gin, K. Y. H., & Ngo, H. H. (2015). Fecal pollution source tracking toolbox for identification, evaluation and characterization of fecal contamination in receiving urban surface waters and groundwater. *The Science of the Total Environment*, *538*, 38–57. PMID:26298247

Urban, F. (2009). Climate-change mitigation revisited: Low-carbon energy transitions for China and India. *Development Policy Review*, *27*(6), 693–715.

Uyterlinde, M. A., Junginger, M., de Vries, H. J., Faaij, A. P. C., & Turkenburg, W. C. (2007). Implications of technological learning on the prospects for renewable energy technologies in Europe. *Energy Policy*, *35*(8), 4072–4087.

Voulvoulis, N., & Georges, K. (2016). Industrial and agricultural sources and pathways of aquatic pollution. In A. McKeown & G. Bugyi (Eds.), *Impact of water pollution on human health and environmental sustainability* (pp. 29–54). Hershey, PA: IGI Global.

Wang, H. F., Sung, M. P., & Hsu, H. W. (2016). Complementarity and substitution of renewable energy in target year energy supply-mix plannin–in the case of Taiwan. *Energy Policy*, *90*, 172–182.

Wang, Q., & Yang, Z. (2016). Industrial water pollution, water environment treatment, and health risks in China. *Environmental Pollution*, *218*, 358–365. PMID:27443951

Wiese, F., Bökenkamp, G., Wingenbach, C., & Hohmeyer, O. (2014). An open source energy system simulation model as an instrument for public participation in the development of strategies for a sustainable future. *Wiley Interdisciplinary Reviews: Energy and Environment*, *3*(5), 490–504.

Wilson, P. (2015). Farm-level actions towards water pollution control: The role of nutrient guidance systems. *Water and Environment Journal: the Journal / the Chartered Institution of Water and Environmental Management*, *29*(1), 88–97.

Yi, H., & Feiock, R. C. (2014). Renewable energy politics: Policy typologies, policy tools, and state deployment of renewables. *Policy Studies Journal: the Journal of the Policy Studies Organization*, *42*(3), 391–415.

Zamfir, A., Colesca, S. E., & Corbos, R. A. (2016). Public policies to support the development of renewable energy in Romania: A review. *Renewable & Sustainable Energy Reviews*, *58*, 87–106.

Zhang, L., & Zheng, X. (2009). Budget structure and pollution control: A cross-country analysis and implications for China. *China & World Economy*, *17*(4), 88–103.

Zhang, X., Pei, W., Deng, W., Du, Y., Qi, Z., & Dong, Z. (2015). Emerging smart grid technology for mitigating global warming. *International Journal of Energy Research*, *39*(13), 1742–1756.

Zheng, D., & Shi, M. (2017). Multiple environmental policies and pollution haven hypothesis: Evidence from China's polluting industries. *Journal of Cleaner Production*, *141*, 295–304.

ADDITIONAL READING

Ambec, S., & Ehlers, L. (2016). Regulation via the polluter-pays principle. *Economic Journal (London)*, *126*(593), 884–906.

Andrews, J., & Shabani, B. (2014). The role of hydrogen in a global sustainable energy strategy. *Wiley Interdisciplinary Reviews: Energy and Environment*, *3*(5), 474–489.

Bakhtyar, B., Saadatian, O., Alghoul, M. A., Ibrahim, Y., & Sopian, K. (2015). Solar electricity market in Malaysia: A review of feed-in tariff policy. *Environmental Progress & Sustainable Energy*, *34*(2), 600–606.

Barbato, L., Centi, G., Iaquaniello, G., Mangiapane, A., & Perathoner, S. (2014). Trading renewable energy by using CO_2: An effective option to mitigate climate change and increase the use of renewable energy sources. *Energy Technology (Weinheim)*, *2*(5), 453–461.

Bosi, S., Desmarchelier, D., & Ragot, L. (2015). Pollution effects on labor supply and growth. *International Journal of Economic Theory*, *11*(4), 371–388.

Corsatea, T. D. (2016). Localised knowledge, local policies and regional innovation activity for renewable energy technologies: Evidence from Italy. *Papers in Regional Science*, *95*(3), 443–466.

Dincer, I., & Acar, C. (2015). A review on clean energy solutions for better sustainability. *International Journal of Energy Research*, *39*(5), 585–606.

Eaton, W. M., Gasteyer, S. P., & Busch, L. (2014). Bioenergy futures: Framing sociotechnical imaginaries in local places. *Rural Sociology*, *79*(2), 227–256.

Grahn, M., & Hansson, J. (2015). Prospects for domestic biofuels for transport in Sweden 2030 based on current production and future plans. *Wiley Interdisciplinary Reviews: Energy and Environment*, *4*(3), 290–306.

Gullberg, A. T., & Bang, G. (2015). Look to Sweden: The making of a new renewable energy support scheme in Norway. *Scandinavian Political Studies*, *38*(1), 95–114.

Heng, L. C., Al-Amin, A. Q., Saidur, R., & Ward, T. A. (2015). Renewable energy choice: Cost and energy analysis of grid connected photovoltaic system in Malaysia. *Environmental Progress & Sustainable Energy*, *34*(3), 866–880.

Jo, C. Y., & White, L. (2013). Polluted air or policy advance in Hong Kong-Guangdong? *Asian Politics & Policy*, *5*(1), 77–106.

Johnson, B. B. (2012). Experience with urban air pollution in Paterson, New Jersey and implications for air pollution communication. *Risk Analysis*, *32*(1), 39–53. PMID:21883333

Jung, T. Y., & Kim, H. J. (2016). A critical review of the renewable portfolio standard in Korea. *International Journal of Energy Research*, *40*(5), 572–578.

Kreifels, N., Mayer, J. N., Burger, B., & Wittwer, C. (2014). Analysis of photovoltaics and wind power in future renewable energy scenarios. *Energy Technology (Weinheim)*, *2*(1), 29–33.

Kulovesi, K. (2014). International trade disputes on renewable energy: Testing ground for the mutual supportiveness of WTO law and climate change law. *Review of European, Comparative & International. Environmental Law (Northwestern School of Law)*, *23*(3), 342–353.

Marconi, D. (2012). Environmental regulation and revealed comparative advantages in Europe: Is China a pollution haven? *Review of International Economics*, *20*(3), 616–635.

Palzer, A., & Henning, H. M. (2014). A future German energy system with a dominating contribution from renewable energies: A holistic model based on hourly simulation. *Energy Technology (Weinheim)*, *2*(1), 13–28.

Park, S. (2015). State renewable energy governance: Policy instruments, markets, or citizens. *The Review of Policy Research*, *32*(3), 273–296.

Pautrel, X. (2012). Pollution, private investment in healthcare, and environmental policy. *The Scandinavian Journal of Economics*, *114*(2), 334–357.

Ploberger, C. (2013). China's environmental issues, a domestic challenge with regional and international implications. *International Journal of Applied Logistics*, *4*(3), 47–61.

Potoski, M., & Prakash, A. (2013). Do voluntary programs reduce pollution? Examining ISO 14001's effectiveness across countries. *Policy Studies Journal: the Journal of the Policy Studies Organization*, *41*(2), 273–294.

Powell, H., & Lee, D. (2014). Modelling spatial variability in concentrations of single pollutants and composite air quality indicators in health effects studies. *Journal of the Royal Statistical Society: Statistics in Society, Series A*, *177*(3), 607–623.

Riesz, J., & Milligan, M. (2015). Designing electricity markets for a high penetration of variable renewables. *Wiley Interdisciplinary Reviews: Energy and Environment*, *4*(3), 279–289.

Ríos, L. G., & Diéguez, J. A. (2016). A big data test-bed for analyzing data generated by an air pollution sensor network. *International Journal of Web Services Research*, *13*(4), 19–35.

Sawhney, A., & Rastogi, R. (2015). Is India specialising in polluting industries? Evidence from US-India bilateral trade. *World Economy*, *38*(2), 360–378.

Smith, M. G., & Urpelainen, J. (2014). The effect of feed-in tariffs on renewable electricity generation: An instrumental variables approach. *Environmental and Resource Economics*, *57*(3), 367–392.

Ward, C. J. (2015). It's an ill wind: The effect of fine particulate air pollution on respiratory hospitalizations. *The Canadian Journal of Economics. Revue Canadienne d'Economique*, *48*(5), 1694–1732.

Wiersma, B., & Devine-Wright, P. (2014). Public engagement with offshore renewable energy: A critical review. *Wiley Interdisciplinary Reviews: Climate Change*, *5*(4), 493–507.

Yanase, A. (2012). Trade and global pollution in dynamic oligopoly with corporate environmentalism. *Review of International Economics*, *20*(5), 924–943.

Zhao, L., & Haruyama, T. (2015). Plant location, wind direction and pollution policy under offshoring. *World Economy*, *38*(1), 151–171.

KEY TERMS AND DEFINITIONS

Climate Change: The long-term shift in the planet's weather patterns or average temperatures.

Contaminant: A substance that contaminates another substance.

Energy: A source of usable power, such as petroleum and coal.

Fossil Fuel: An organic substance from the geologic period that is found underground and is used as a source of energy.

Global Warming: The gradual heating of Earth's surface, oceans, and atmosphere.

Greenhouse Gas: Any gaseous compound in the atmosphere that is capable of absorbing infrared radiation, thus trapping and holding heat in the atmosphere.

Pollutant: Something that pollutes, especially a waste material that contaminates air, soil, and water.

Pollution: Something introduced into the environment that is unclean or has a harmful effect.

Renewable Energy: Energy that is generated from natural processes that are continuously replenished.

Chapter 6
Pollution Exposure to Humans and Its Assessment

Rajmal Jat
Indian Institute of Technology Roorkee, India

Veerendra Sahu
Indian Institute of Technology Roorkee, India

Bhola Ram Gurjar
Indian Institute of Technology Roorkee, India

ABSTRACT

Exposure analysis is the receptor-oriented approach of the pollution-level measurement. In this chapter, a detailed discussion is provided of the fundamentals of exposure analysis, methods of measurement, basics of models used for the prediction of pollution concentration indoors and outdoors, and a brief discussion about the health impact of selected pollutants. A detail of fundamental of indoor air quality (IAQ) models like mass balance and CFD models is discussed. Also, basic structures of community multiscale air quality model (CMAQ) and AIRMOD ambient air dispersion models are described. It is observed that measurement of pollution exposure by direct method requires more time and effort as compared with the integrated exposure and stationary measurement. AIRMODE is steady state model and based upon the Gaussian dispersion model. CMAQ is capable of simulating the pollution level for the range of geographic scale for multiple pollutants.

INTRODUCTION

Unlike the traditional approach of environmental science which is the source oriented approach, exposure science has receptor oriented approach. In the receptor oriented approach, level of pollution which is reaching to the target is measured

DOI: 10.4018/978-1-5225-3379-5.ch006

using personal monitors and backward analysis is carried out to find the sources of pollutants. Exposure to the pollutants is defined with respect to the target of the risk. If the objective is to protect the ecological system, the target may be some organism in the ecosystem, whereas, if it is to protect public health, the target will be human being. As human being is considered as the most important component of pollution system, exposure is generally referred to the human exposure and human being is treated as the receptor of environmental pollution (Ott, 1982). Stationary measurement of pollutant in outdoor is used to represent the exposure but doesn't represent the actual personal exposure because majority of people spend more than 80% time in indoors (Spengler & Sexton, 1983). Therefore, now a day's indoor air pollution studies and personal exposure studies gaining attention of the scientist all over the world.

There is an abundant epidemiological evidence suggesting association between increased air pollution and adverse human health (Dockery et al., 1993). An empirical relationship between adverse respiratory and cardiovascular effect and ambient suspended particulate matter (PM) concentrations has been observed by various research studies in the past (Vedal, 1997). Various studies in the past have shown that outdoor particulate matter (PM) concentration is in reality a poor indicator of both indoor and personal particulate exposure level (Clayton et al., 1993). Recent progress in personal exposure and health assessment has led to an increased interest in indoor air pollution. Quantifying human exposure to air pollution is useful for risk assessment and carried out through measurement methods and models. In this chapter description of exposure, measurement methods, various models and health effect of pollutant will be discussed.

BACKGROUND

According to the risk assessment model, harmful health effects caused by the various pollutants in the environment can be assessed. Risk assessment is the four step process involves Hazard Identification, Exposure assessment, Dose response assessment and Risk characterization process. In the first step of hazard identification of risk assessment, contaminants which are responsible for the human health hazard are identified, and their qualitative analysis with respect to the health effect is presented. After the hazard identification, consideration in the risk assessment is given to the exposure assessment and dose response assessment. In the dose response assessment, human being is exposed to the different known values of contaminant concentration and its corresponding health effect at each level is analyzed. Further, Exposure assessment for the population is carried out using various available methods and the result of that is coupled with the dose response analysis. By doing so the risk characterization to the individual can carried out. A flow diagram for the above stated process risk assessment is presented in the figure 1.

Since the exposure assessment is the integral part of the risk assessment process, measurement and modeling related with the exposure assessment becomes very essential. Exposure assessment model can predict the exposure even though few measuring data are available. These models are very useful in case where obtaining the measurement data involves very high cost.

Relation between the target and the agent can be established with the full knowledge of the five component of risk model which have the series of connection as shown in figure 2 (Akland et al., 1984). Each component in this model is dependent on the one preceding it such that the output from the one component becomes the input for the next. Lack of information about any of the component of this full risk model makes

Figure 1. Risk Assessment Model (source: EPA, 1992)

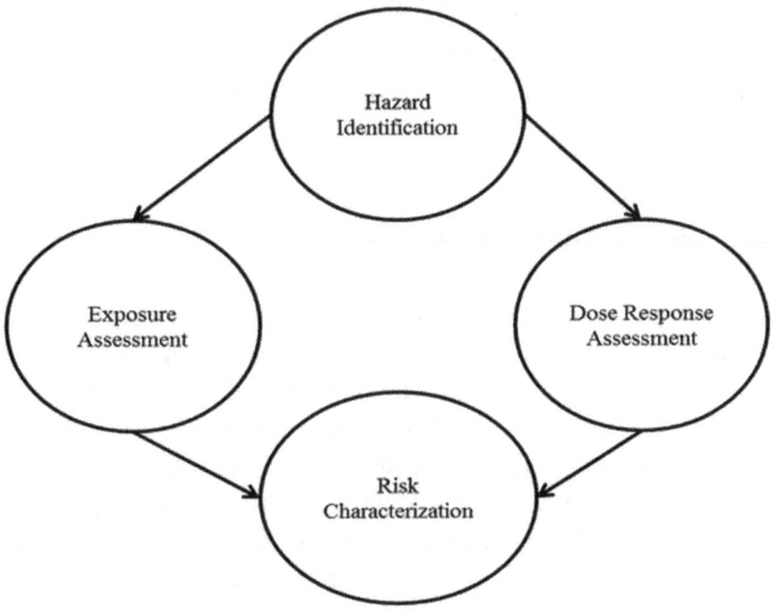

Figure 2. Risk Model (source: Akland et al., 1984)

the characterization of relationship between source and the effect difficult. Therefore, knowledge of every component is required for fully established relationship between source and the effect. Pollutant is released into the atmosphere form its source and gets into the environment where it is transported into the various media like air, water, soil and surface. Apart from the transportation it also undergoes into various physical and chemical transformations and converted into the secondary species. After the transformation process, pollutants have their different concentration into the different media, and when the human being comes in contact to the any of the media it is subjected to the dose of pollutant as per the concentration. This dose may be resulted into the human health effect.

The total human exposure brings the concept of exposure to the pollutant through the all possible routes. To estimate the total human exposure, contact boundary and concentration in each of the medium should be fully understood. Some pollutant may reach to the human being through only one route like O_3 can reach only through air. Whereas, some may reach through multiple routes like air, water, food and so forth. Therefore, all possible route of the particular pollutant must be known to represent the human exposure. There are several methods of exposure assessment under the head of direct assessment and indirect assessment which are discussed in detail in this chapter.

SOME IMPORTANT DEFINITIONS

Definition of exposure has been given by many researchers in different frame works. It is defined as the contact between an agent and a target at contact boundary over the exposure period (Ott et al., 2006). Exposure in general defined as the time integrated and time averaged. Always exposure is confused with the dose of pollutants. Dose is defined as the amount of the pollutant that gets into the receptor in a certain period of time, crossing the contact boundary. Further as per the type of boundary the dose type is attributed. For example, if the contact boundary is absorption type the dose will be called as absorbed dose. Since we are confined our discussion around the exposure to the human being, hereafter the exposure will be referred as human exposure.

- **Agent:** Agent is defined as the chemical, physical or biological entity which may cause the harmful effect after contacting the target. Not all the agents are harmful. As per the physical forms agents are of the two types: energy form agents and matter from agents. Energy form agents include light, sound, heat, radiation, magnetism, etc. whereas matter form agent consists chemical mass, microorganism, particle matter etc. Air, water and solid are the medium for transporting the agent to the receptors. For matter form agent amount is represented as mass per unit volume also called as concentration and for energy form presented as amount per unit area (i.e. intensity) (Ott et al., 2006).

- **Targets:** All physical, biological, and ecological entities come under the category of targets. Like humans, animals, plants, walls and houses are the few examples of target. Before conducting the exposure studies target and agent should be clearly defined and this selection depends upon the agent of interest (Ott et al., 2006). For example exposure of volatile organic compound (VOC's) to the human being. Sometime exposure route is confused with the exposure pathway. Exposure route is the way of penetrating through the contact boundary and exposure pathway is the path from the source of pollutant to the receptor.
- **Exposure:** It is defined as an event when source comes in contact with the target, and if the target is the human being than it is called as the human exposure (Ott et al., 2006). It has the three major components exposure surface, contact volume and concentration of the agent. Exposure surface is defined as the area on the target which is exposed to the agent of interest. Also, at the same time the concentration of agent will vary from the point to point over the exposure surface. Contact volume is referred to the volume adjacent to the exposure surface with the certain exposure concentration. Exposure concentration is the mass per unit volume in the vicinity of exposure surface, and it can vary spatially and temporally. An average spatial exposure can be defined mathematically from the following equation.

$$E_{(x,y,z,t)} = \frac{\iiint C_{(x,y,z,t)} dxdydz}{\iiint dxdydz}$$

where;

$E_{(x,y,z,t)}$ = Average spatial exposure at time 't'

$C_{(x,y,z,t)}$ = Point exposure in the contact volume (dxdydz) at time 't'

Another important aspect of the agent concentration is the time integrated exposure, integration of average spatial exposure over the exposure duration. Exposure duration is the time interval of contact between agent and the target (Ott et al., 2006). The same is depicted with the following mathematical equation.

$$I_{\Delta t} = \int_{t_1}^{t_2} E_{(x,y,z,t)} dt$$

where;
$I_{(\Delta t)}$ = Time Integrated exposure over the exposure duration 'Δt'
Δt= Exposure duration (t_2-t_1)

If the time integrated exposure is divided by the exposure duration it will be called as the time averaged exposure.

$$I_{ta} = \frac{\int_{t_1}^{t_2} E_{(x,y,z,t)} dt}{t_2 - t_1}$$

where $I_{(ta)}$ is the time averaged exposure.

- **Dose:** It is the quantity of agent which enters the agent after crossing the exposure surface in a particular duration. There is the close link between exposure and dose. It is possible to do not have the dose even the exposure present but without the exposure dose is not possible (Ott et al., 2006). For example, if a human being is exposed to the chemical over the skin and that chemical is unable to penetrate through then it is said that there is exposure but without dose. The instantaneous dose rate is defined as the amount of the agent passes through the contact boundary per unit area per unit time, and if it is integrated over the exposed area and duration of the time it will provide the total amount of agent entered into the body. Mainly dose is classified as the intake dose and absorbed dose. If the pollutant crosses the contact boundary without diffusing through the resisting boundary layer it is called as the intake dose. The example of intake dose is through inhalation and ingestion. Whereas, if the pollutant diffused and crosses the contact boundary it is called as the absorbed dose. Some example of resisting boundaries over which pollutant diffused are skin and respiratory tract lining.

METHODS OF MEASUREMENTS

Exposure measurement can be carried out in two ways: direct method and indirect method. Direct method is associated with the use of personal sampler and individual level human exposure is determined using panel studies. Biomarkers are also comes under the head of direct methods. In indirect methods, either human exposure is measured by stationary sampling at some representative sites or use of suitable model is deployed for human exposure measurement (Ott, 1982; Lioy, 1995). There are several factors which affect the selection of the method like sensitivity, accuracy, precision, threshold limit, cost and applicability.

Direct Method

- **Personal Sampling:** In the direct method of sampling passive samplers are used. There are number of passive samplers available for measurement of gaseous pollutants and particulate matter. In general for gaseous pollutants their design includes badge, diffused tube-filter absorbent surface, or diffused tube with packed solid adsorbent. Pollutant concentration is determined by optical method, or by extraction into the solution which further quantify the pollutant concentration. Passive portable sampler is attached to the individual subjects at the breathing level such that the exposure in the close proximity of the individual can only be observed. Passive sampler are light, battery operated, and exists as small personal badges. These instruments can also be used for stationary sampling in indoor and outdoor environment. Passive samplers are available for particulate matter of various size range, VOC's and co-pollutant gaseous (Louie & Pierce, 1988). There is several personal exposure studies have been conducted using this technique. Wang C., et al. (2017), conducted exposure study on the panel of 36 nonsmoking healthy college students in shanghais, china during December, 17, 2014 to July 11, 2015 using MicroPEM monitors (v3.2MicroPEM,RTI,USA) to record the real $PM_{2.5}$ concentration and HOBO data logger for monitoring personal temperature and relative humidity. This study concluded that exposure to $PM_{2.5}$ may decrease the lung functioning. Similarly, Devi J.J. et al. (2012), conducted exposure study on a panel of 18 healthy students in the Indian institute of technology Kanpur, India. Portable instruments: Optical particle counter (OPC; model 1.108, Grimm), Multiple gas monitor (IAQRAE Gas Monitor, Model PGM-5210; RAE Systems Inc.), and Condensation particle counter (CPC model 3007, TSI Inc.) were used for the measurement of particulate matter and pollutant gaseous.

- **Biomarkers:** Since many years biomarkers are being utilized in the occupational studies and now days gaining attention in the environmental studies. It is the chemical compound which is found in the body tissues. Biomarkers which are having first existence in the environment and then enter into the body termed as the parent compounds. Some compound in the body get metabolize from the parent compound may be used as the biomarkers as well. Mainly biomarkers are classified into the two categories; biomarker of exposure and biomarker of effect. To identify the occurrence of exposure and the sometime route of exposure, biomarker of exposures are used. In contrast to the biomarker of exposure, biomarkers of effect are the changes in the body caused by the exposure. Though a classification of such types is given but a clear demarcation between these two are always difficult. While relating the biomarker of exposure with a particular route of exposure for quantitative estimation of exposure, there must be only on way of exposure to enter into the body. With the multiple ways of entry into the body biomarker of exposure doesn't provide the quantitative estimation of exposure. For example, chloroform has the multiple ways to enter into the body namely: via inhalation, through drinking water, or adsorption through skin. So in that situation quantitative estimation of exposure cannot be provided by the biomarker. Similar type of problem in quantitative estimation happens when the biomarker is originated from more than one chemical. Since the objective of the use of biomarker is to get the quantitative exposure, a relationship between level of biomarker and level of exposure must be known. In general breath, blood, bone, teeth, fat, hair, urine, nail and feces are used for the biomarkers. Most of the pollutants are investigated by biomarkers like volatile organic compound (VOC), pesticide, polyaromatic hydrocarbons (PAHs), furan, dioxins and some metals (Ott et al., 2006).

VOC's like benzene, toluene, styrene and tetrachloroethylene are considered as the parent compound biomarkers of exposure. Krotoszynskik, Bruneau and O'Neill (1979), conducted a study in which breath samples are taken for targeting many VOC's. Formaldehyde is used in the presses wood, stairs, particleboard and major components in home, and found human carcinogen. It's exposure and uptake is studied using a metabolite in urine. For biomarkers of metals like arsenic, fingernails and hair tissues are used because they form complexes with fibrous protein in hair and nail (Mandal, Ogra, and Suzuki 2003).Exposure to the lead is estimated using the blood. Since many years lead in blood is being used as the biomarker and reduced level of it represent the success of the environmental action taken, especially removal of lead from the gasoline. Researcher revealed that the methylmercury can be passed from the pregnant women to her children which reduce the growth of the brain (Ott et al., 2006). Level of methylmercury in the pregnant women is determined by the hair sample (Cernichiari et al. 1995). Yanagisawa et al. (1988), studied Exposure to NO_2 and tobacco smoke using hydroxyl-proline and creatinine ration as the biomarker in urinary samples. This ratio has the direct relation with the exposure to NO_2.

Indirect Method

Outdoor Sampling: In almost every epidemiological study measurement by the ambient sampling network is considered as the representative of pollution level for the given area. In contrast to the direct method, this method of exposure measurement represents a single value for the population. A network of stationary samplers, in general, is laid down within each country. Sometimes these are equipped with the online monitor to give the real time data at a desired time intervals (Louie & Pierce, 1988).

Integrated exposure: In this approach microenvironment model is used. Microenvironments are defined as the air space with homogenous pollution concentrations such as room, mess, restaurant, kitchen, offices etc. (Duan, 1982). Measurement of pollutant is carried out in each microenvironment and with the help of time activity, time which an individual spend in the various microenvironments is observed. Further with the help of time averaged exposure equation integrated exposure is computed. More specifically if we write the time averaged exposure equation in the discrete form it can be understood more clearly. Following equation is the discrete form of the equation where C_j is the concentration of pollutant in the jth microenvironment, T_j is the time which is spent by the individual in the jth microenvironment and E is the integrated exposure.

$$E = \frac{\sum_{j=1}^{n} C_j T_j}{\sum_{j=1}^{n} T_j}$$

E→ Time weighted average exposure
C_j→ Average concentration of pollutants in 'j^{th}' microenvironment.
T_j→ Average time spent by individual in 'j^{th}' micro environment.

Figure 3 is graphical representation of integrated personal exposure measurement. Where different source contributing pollutant in different microenvironments and the time spend by the individual in each microenvironment is observed by maintaining time activity diary or questioner. With the know pollutant concentration in various

Figure 3. Graphical representation of integrated exposure (Source: Monn, C., 2001)

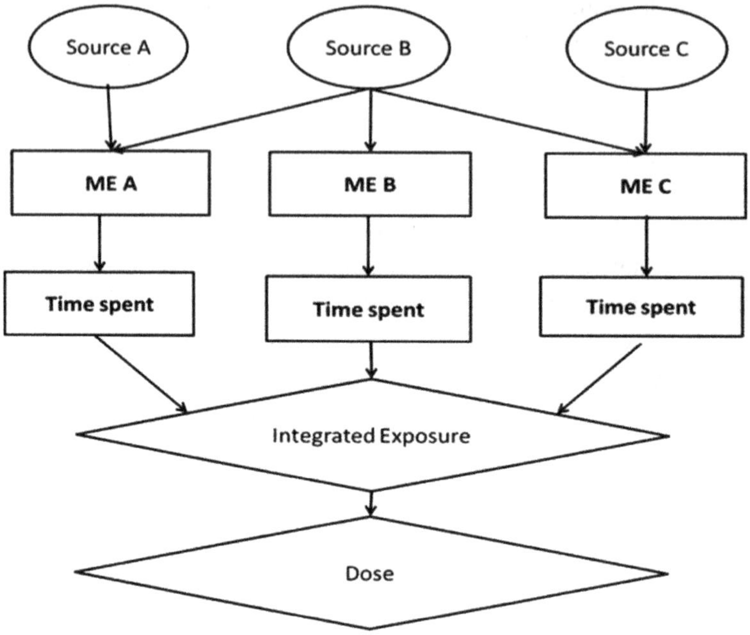

microenvironment and time spent by individual, the integrated human exposure can be computed. In this type of integrated personal exposure estimate there is a good chance of losing the information on the magnitude of peak concentration and their time of occurrence within the total time period.

An example of indirect approach will make this clearer. Let's say exposure to $PM_{1.0}$ number concentration (#/cm^3), denoted by 'c', into different microenvironments inside an institute campus and duration (hrs/week), denoted by t, are known and we need to compute the integrated exposure. Computation with detail information of time weighted average exposure (E) is given in the table 1.

MODELS

Air pollution modeling is categories into the two parts: Indoor air quality modeling and atmospheric dispersion modeling. It ranges through empirical, deterministic and semi-empirical models (Louie & Pierce, 1988). In the following section a brief introduction about the fundamental of these two modeling approach is discussed.

Table 1. Exposure to $PM_{1.0}$ (#/cm³)

Microenvironments	t (hours/week)	c (#/cm³)	c × t
Hostel Room	88.3	13318	1175979.4
Mess	7	72237	505659
Laboratories	21.1	16729	352981.9
Lecture halls	14.8	4428	65534.4
Computer center	3.2	997	3190.4
Library	3.8	6155	23389
Faculty building	2.6	7865	20449
Bank	0.6	14598	8758.8
Health center	0.1	8936	893.6
	$\sum t = 141.1$		$\sum c \times t = 2156835.5$
	$E = \sum c \times t / \sum t = 15286$ (#/cm³)		

Atmospheric Dispersion Models

A model is the generalized picture of complex real world phenomenon. Degree of generalization depends upon the knowledge and technology available in hand and problem of interest to be solved. By modeling we are able to provide the solution for any form of environmental issue. If the real world scenario is scaled down and presented, it is termed as the physical model. Whereas, a mathematical formulation of physical problem is called as the mathematical model. Atmospheric dispersion models are the mathematical type models and capable of simulating the physical and chemical phenomenon associated with the pollutants in the real world situation. Dispersion models can be presented either in forms of tables, chart and formulae on paper or as computer programmer with graphical user interface. Figure 4 represent the approach of dispersion modeling. Atmospheric dispersion models required the input of emission inventory, meteorological data, geographical data, and initial

Figure 4. General dispersion modeling approach (Source: Louie & Pierce, 1988)

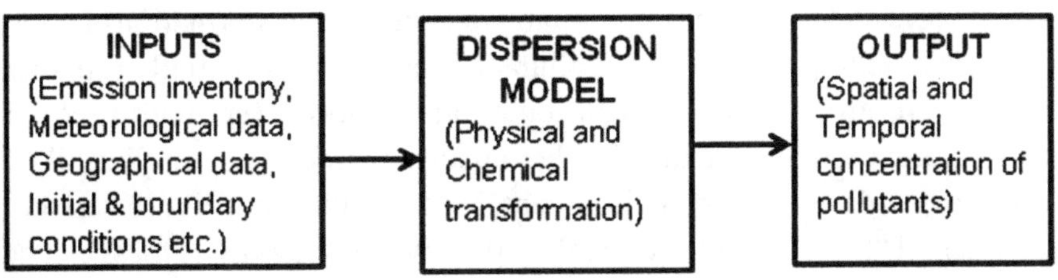

boundary conditions to the model. Thereafter, physical and chemical transformation model inside the dispersion model does computation on the given inputs and produce output in term of spatial and temporal distribution of pollutant concentration.

Indoor Air Quality Models

Indoor air quality (IAQ) models offer estimates of indoor pollutant concentrations derived from outdoor and/or internal sources. IAQ models estimate/predict indoor concentrations of contaminants as a function of outdoor pollutants concentration, air-exchange rates and contaminants sources and sink in indoor. IAQ modeling is always preferred over the expensive and invasive large field studies, in this context a range of statistical, mathematical and computational models have been developed, ranging from simple regression model to more complex deterministic models, single zone to multi zone, microscopic to macroscopic, and static to dynamic models.

In general IAQ models can be classified in two categories as statistical models and mathematical models. Further mathematical models can be categorize as microscopic and macroscopic models, microscopic models includes IAQ models which are based on mass balance equation, whereas macroscopic models contains computational fluid dynamic (CFD) models (Goyal, & Khare, 2010).

Statistical model: In the statistical model, outdoor pollutant concentration is related with the indoor concentration using regression analysis. The coefficients of regression are computed based upon the experimental data using least-square techniques. General form of regression equation used for indoor-outdoor air quality is as follows.

$$Y = a + bX$$

where;
Y = Indoor concentration of respective pollutant species
X = Out pollutant concentration of respective pollutant species
a & b = Regression coefficient

Mass balance models: Mass balance models are deterministic in nature, developed through the basic understanding of pollutant transfer and transformation phenomena. These are capable to predict indoor pollutant concentration spatially (i.e. different buildings and rooms) as well as with temporal variations (i.e. with time). These models are often more useful and simple in nature with the assumptions like study state condition and well mixed environment. Mass balance models are derived from the fundamental "law of conservation of matter" which states that "mass can neither be created nor destroyed, though it may be rearranged in space, and altered in form"

(Indoor Pollutant, 1981). Consider the room or kitchen as the control volume as presented in the Figure 5.

The basic mass balance equation can be written as:

Accumulation of pollutant concentration in control volume = Pollutant concentration generated indoor + Pollutant concentration entering from outdoor - Pollutant concentration leaving from control volume

Mass balance model can be expressed as an ordinary differential equation:

$$V\frac{dc}{dt} = G + QC_O - QC \quad (1)$$

where;
V = Volume of the control volume
G = generation rate of pollutant (mass/time)
C_0 = Outdoor pollutant concentration (mass/volume)
C = Pollutant concentration inside the volume (mass/volume)
Q = Air Infiltration rate

on integration of the equation 1

$$\int_{C_i}^{C} \frac{dc}{G + QC_O - QC} = \int_0^t \frac{dt}{V} \quad (2)$$

where C_i is initial pollutant concentration of control volume

Solving the equation 2 for C

$$C = \left(C_O + \frac{G}{C}\right) + \left(C_i - \frac{G}{Q} - C_O\right)e^{-\lambda t} \quad (3)$$

Figure 5. Simple Box Model for Indoor air modeling

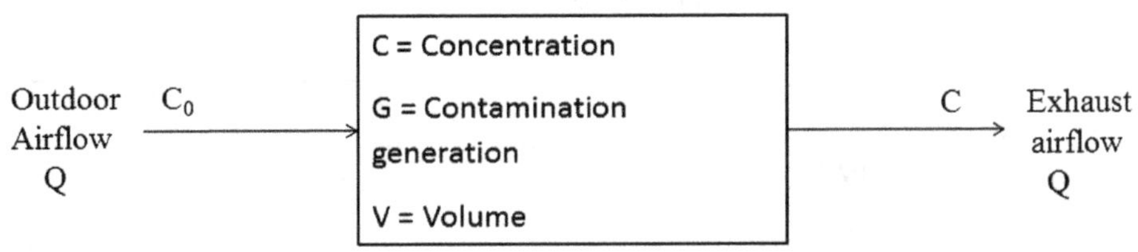

where;

t = time

λ = Outdoor air exchange rate, specified in air change per hour (ach)

λ = Q/V

Equation 1 i.e. statement of the mass balance model, can be more complex depending upon the complexity of the problem which may include the range of different factors like ventilation condition, air mixing, Indoor air chemistry, sedimentation, deposition of pollutants etc., which affects the concentration of indoor contaminants.

For example, consider a mechanically ventilated room of volume V as shown in the figure 6, in which fresh air enters from the outside and passes through a filter at a rate q_0. Some amount of the indoor air is recirculated through additional filter at a rate q_1, and air infiltrates in the control volume at a rate q_2. Each filter is characterized by a factor $F \cong \left(C_{inlet} - C_{outlet}\right) / C_{inlet}$., the contaminant concentration is assumed to be uniform throughout the control volume. C and C_0 represents the contaminant concentration in indoor and outdoor, respectively at given time'ť'. 'S' indicates the rate of pollutant added to indoor, and 'R' represent the rate at which pollutant removed through sink.

The pertinent equation can be written is:

$$V\frac{dC}{dt} = q_0 C_0 \left(1 - F_0\right) + q_1 C \left(1 - F_1\right) + qC_0 - \left(Q_0 + q_1 + q_2\right)C + S - R$$

Input rate due torecirculated + infiltratedOutput + SourceSink
makeup air air air rate rate rate

(4)

Figure 6. Mechanical ventilated room of volume V.

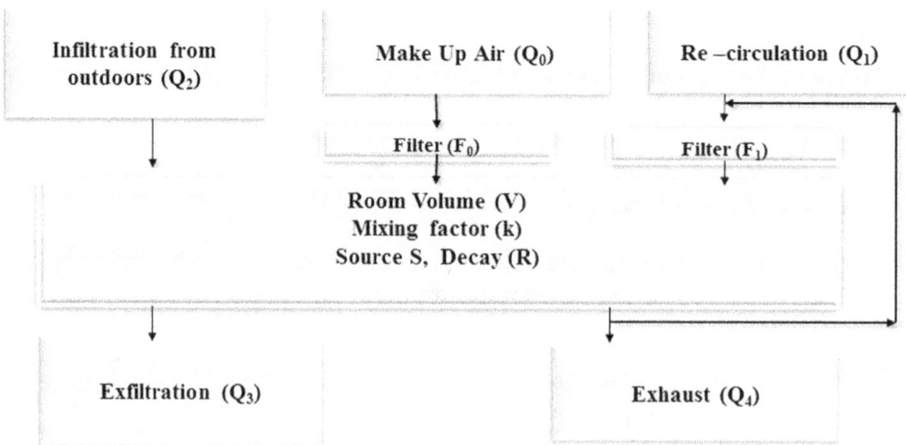

The decay rate is a function of C. yet, in indoor air quality modeling, the sink rate is often assumed constant. As shown in Equation 4, results of mass-balance model consistently involve parameters that must be assessed autonomously. Design parameters, like volumes and surface areas, can be measured straight or found from drawing of the buildings. Generally the ventilation parameters and pollutant rates (both generated and release) are difficult to determine, such uncertainty and difficulty of various parameter leads the mass balance model to the compartment model. Usually, a region or volume can be considered as compartment only if the pollutant concentration of the region/volume is uniform throughout the time of interest.

Given the availability of required information, a mass balance model, can be applied as single compartment model or multi compartment model. The simplest method to mass balance modelling is provided by the single compartment model, which assumes that uniform mixing inside the compartment and there is no spatial variation of pollutant within the compartment (Zierler, 1981).whereas multi compartment model is more complex where the building or region comprises of several well-mixed zones, in such case the pollutants as not uniformly distributed in the compartment and rate of mixing is low compared to pollutant resident time.

Due to the difference in building characteristics, pollutant properties and factor affecting IAQ of the building, no single mass balance model existed which can cover all conditions (Nazaroff, 2004). Hence a number of studies carried out and the different mass balance IAQ model has developed to predict indoor air quality of buildings. Imperial College London has developed INDEX Indoor Exposure Model, based on a comprehensive review of earlier research experiment and observation based on mass balance studies. INDEX is a deterministic model which estimates indoor pollutant concentrations and indoor/outdoor concentration ratios for a demonstrative room. Other indoor air quality multi-zone model like CONTAM Multizone Airflow and Contaminant Transport Analysis Software, has been developed by the American National Institute of Standards and Technology (NIST). CONTAM is a computer program which predict chemical pollutant concentration, personal exposure to chemicals and air flow of the building of the interest. Whereas USEPA has developed cost effective IAQ model called "RISK" which is designed to calculate individual exposure from indoor air pollutant and the model also allows to assess the effects of air cleaner on IAQ and exposure. The American Society of Heating, Refrigerating, and Air-Conditioning Engineers (ASHRAE) and the Air Infiltration and Ventilation Centre (AIVC) are some more examples of organizations which provide the resources for ventilation modeling (IEHIAS, website).

- **Macroscopic Models:** Macroscopic models of IAQ are mainly based on Computational fluid dynamics (CFD) techniques; The CFD technique has developed a progressively widespread method to assess indoor air quality of simple to most complex types of buildings. CFD is a powerful tool which permits the user to simulate airflow, thermal comfort and indoor contaminants distributions by varying the different parameters that affect the concentration and flow of indoor contaminants (Pitarma at el., 2004). CFD models are computational models that comprises the solution of the set of non-linear partial differential equations, which govern the physics of the flow i.e. conservation of mass, momentum and energy.

Due to the restrictions of the experiments and the growth in the computational power and affordability of computers, CFD offers a practical choice for computing the airflow and contaminants distributions in indoor environments. This is a more real approach to divide the region/volume of interest into a number of unreal sub-volumes, or elements. These sub-volumes typically do not have solid boundaries; rather, these allow gases to pass through their bounding surfaces (Spengler et al., 2001).

In general all practical flows have some degree of random turbulence due the inertial and viscous forces of the fluids and the transport of momentum, energy and pollutants is affected by such turbulent instabilities. Therefore turbulence must be incorporated in formulation and solution of the equations of motion, for more accurate results of CFD models.

The correctness of CFD models is very sensitive to the boundary conditions and the grid set-up. Normally, the boundary conditions for the CFD models consist of the inlet and outlet characteristics, enclosure surfaces characteristics and internal substances. The temperature, velocity and turbulence of the air incoming through diffusers or windows govern the inlet conditions, whereas the interior surface temperatures, vapour pressure and humidity are the indispensable boundary conditions for the region of interest (Spengler et al., 2001).

Governing Equations of IAQ CFD Model

Conservation of Mass

$$\frac{\partial}{\partial x}(\rho \mathcal{U}) + \frac{\partial}{\partial y}(\rho \mathcal{V}) + \frac{\partial}{\partial z}(\rho \mathcal{W}) = 0 \tag{5}$$

Conservation of Momentum

$$\frac{\partial}{\partial x}(\rho \mathcal{U}) + \frac{\partial}{\partial y}(\rho \mathcal{U}\mathcal{U}) + \frac{\partial}{\partial z}(\rho \mathcal{U}\mathcal{V}) + \frac{\partial}{\partial z}(\rho \mathcal{U}\mathcal{W}) = -\frac{\partial p}{\partial x} + \frac{\partial}{\partial x_j}[\mu\left(\frac{\partial \mathcal{U}}{\partial x_j} + \frac{\partial \mathcal{U}_j}{\partial x}\right) \tag{6}$$

Conservation of Energy

$$\frac{\partial}{\partial t}\left(\rho C_P T\right)+\frac{\partial}{\partial x}\left(\rho C_P \mathcal{U} T\right)+\frac{\partial}{\partial y}\left(\rho C_P \mathcal{V} T\right)+\frac{\partial}{\partial z}\left(\rho C_P \mathcal{W} T\right)=\frac{\partial}{\partial x_j}\left(k\frac{\partial T}{\partial x_j}\right)+q \quad (7)$$

Conservation of Contaminants

$$\frac{\partial C}{\partial t}+\frac{\partial}{\partial x}\left(\mathcal{U}C\right)+\frac{\partial}{\partial y}\left(\mathcal{V}C\right)+\frac{\partial}{\partial z}\left(\mathcal{W}C\right)=\frac{\partial}{\partial x_j}\left(D\frac{\partial C}{\partial x_j}\right)+S \quad (8)$$

where:
u,v& w= air velocity in the x,y and z direction respectively in m/s
ρ = air density in kg/m^3
μ = air viscosity Pa.s
β = thermal expansion coefficient of air in K^{-1}
g = gravitational acceleration in m/s^2
t = time in sec.
p = pressure in Pa
T = temperature in K
T_∞ = reference temperature in K
Cp = air specific heat in J/kg K
k = air conductivity in W/m k
C = concentration of contaminant in kg/m^3
D = molecular diffusion coefficient for the contaminant in m^2 /s
S = volumetric contaminant generation rate in kg/m^3 s
q = heat within the control volume in W/m^3

Equations 5 to 8 fully illustrate the transient fluid motion, heat and pollutant transfer of the control volume. The energy and concentration equations have identical structures to the momentum equation. All equations comprises of transient, convection, diffusion and source terms .The numerical model comprises the discretization of the flow problem, the boundary conditions and the turbulence modelling. Here validation of model is defined as the experimental verification. There are six unknowns (temperature, pressure, concentration and three velocity components) in these six equations, therefore the problem is closed. The analytical solution of a three dimensional coupled, non-linear, turbulent, partial differential equations, is not possible. Thus the calculation of such fluid problem needs to be discretized into space and time using either finite difference or finite element methods. The region of interest is divided into a large number of minor cells, called as the grid and the equations are solved through the second order interpolation scheme (Pepper & Carrington, 2009).

REVIEW OF AIR QUALITY MODELS

There is plethora of air dispersion models are available, specially developed by united state environmental protection agency (USEPA). Out of which, Community Multiscale Air Quality (CMAQ) system and AIRMOD are the most popular models for dispersion modeling of pollutant into the atmosphere. In this part a brief description of these two models is provided. CMAQ can address the issues related with the multiple pollutants and at different scales (i.e. urban scale to regional scale) (Byun & Schere, 2006). It uses generalized coordinate system and scale dependent dynamics of the atmosphere. This system is the combination of three components emission modeling, meteorological modeling and chemical transportation modeling. It also provides the facility of sensitivity analysis for output concentration of pollutant with respect to the individual physical and chemical atmospheric process. AIRMOD is the steady state plume dispersion model based on the planetary boundary layer turbulence. Modeling system of AERMOD includes AIRMOD as main program and AERMET and AERMAP as preprocessors (Cimorelli, 2004). AERMET is used to generate the planetary boundary layer (PBL) parameters and AERMAP for generating terrain and receptor data using digital elevation data. A detail of these two models are given below.

Community Multi Scale Air Quality Modeling (CMAQ)

CMAQ is a multipurpose tool which has mainly three major components Emission modeling system, Meteorological modeling system and Chemical transport modeling system such that output of emission and meteorological model and several other data source through suitable interface incorporated into the chemical transport model (Byun & Schere, 2006). It can do modeling for multiple pollutants and at multiple scales, ranging from regional to urban scale. CMAQ model provide flexibility of substituting the other emission and meteorological models. It can simulate the transport, transformation and deposition of pollutants ozone, particulate matter, airborne toxics, and acidic and nutrient pollutant species (Byun, & Schere, 2006).

Figure 7. Flow diagram of CMAQ modeling system (Source: Ching & Byun, 1999)

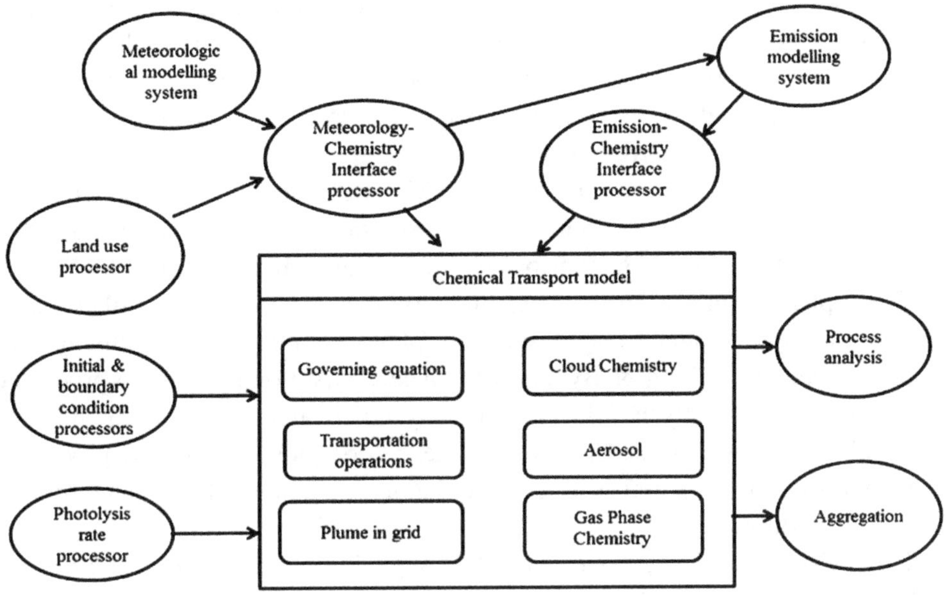

Figure 7 is showing the structure of CMAQ. For the Models-03, meteorological field is generated by mesoscale model (MM5), a product of Penn State University (PSU)/ National Center for Atmospheric Research (NCAR), and emission is generated by Emission Project and Processing System (MEPPS). Processing of emission inventory data, future projection and pre-processing of data for use in the CMAQ are the primary function of the MEEPS. It is created on the Geocoded Emission Modeling and Projection System (GEMAP) and gives emission concentration for each species which is associated with the CB-IV or RADM2 chemistry mechanisms (Ching J. & Byun D. 1999). Interface processors are used by the CMAQ for processing the input data from emission and meteorological systems and other processors for photolysis rates and initial boundary condition development. For translating the data from MEPPS emission model to use in chemical transport model (CCTM), Emission chemistry interface processor (ECIP) is used which generated hourly three dimensional emission data for mobile, area and point sources. Interface processor for meteorology model is employed in form of Meteorology Chemistry Interface Processor (MCIP) which provides many functions like meteorological data interpolation, coordinate system transformation, parameters computation for cloud cover and planetary boundary layer (PBL). Interface of Initial conditions and boundary conditions (ICON and

BCON) is provided to generate concentration fields for individual chemical species for the beginning of a simulation and the grids surrounding the modeling domain. Photolysis rates are produced using photolysis processor (JPROC) which is required as an input for CMAQ chemical transport model (CCTM). Temporal variation of photolysis rate depends upon the vertical ozone profile, temperature profile, aerosol number density and the earth albedo.

CMAQ chemical transport model (CCTM) is the heart of the model which simulates the atmospheric chemistry, transportation and deposition phenomenon into the atmosphere. CCTM has various science options which include RADM2 and CB-IV gas phase chemistry mechanics, suitable numerical solvers, particulate matter estimation algorithm, plume-in-grid approach, photolysis rates and multidimensional advection schemes (Ching & Byun, 1999). The CMAQ output can be processed to provide process analysis information analyzed further to provide aggregated statistical information. Since CCTM simulate the complex atmospheric process, it is required to perform the post processing the output generated for the credibility and understanding.

Sensitivity analysis is carried out in CMAQ using a tool called as the process analysis. Process analysis helps in finding the relative influence of various physical and chemical processes on pollutant concentration and can provide a deep insight of significance of one process over the other. Also, for seasonal and annual estimate of pollutant, a statistical procedure called as 'aggregation' is employed by CMAQ.

AIRMOD

The AERMOD dispersion modeling tool is a steady state model. It is applicable to a range of geographical conditions like rural and urban area, flat and complex terrain; also it incorporates the all types of possible sources of pollution such as point source, area source and volume source (Cimorelli, 2004). While applying for the stable boundary layer (SBL), Gaussian distribution of concentration in both vertical and horizontal direction is considered. For convective boundary layer (CBL), bi-Gaussian probability density function is used in vertical direction and horizontal distribution remains similar to the case of SBL. AERMOD is capable of characterizing the PBL. It takes minimum observed data of meteorological parameter to construct the vertical profile of various parameter like wind speed, wind direction, turbulence, temperature and temperature gradient. Extrapolation of observed data is carried out in AERMOD using similarity theory. It considers the plume of lofting type and uses latest concepts for modeling the flow and dispersion in the complex terrain. Observation regarding cloud cover is also incorporated in AERMODE as it plays vital role in the stability of atmosphere. AERMOD is capable of taking into consideration the inhomogeneous nature of PBL. Average of parameters of actual PBL is carried out and used as effective parameters for the homogeneous PBL.

Figure 8 is showing the modeling system of AERMOD which comprises the AERMOD as main program and two pre-processor AERMET and AERMAP for pre-processing and generating compatible PBL parameters and terrain data respectively. The main objective of AERMET is to calculate the boundary layer parameters using observed surface characteristics and meteorology which are used by the meteorological interface inside the AERMOD to generate the profile of necessary meteorological variable.

HEALTH EFFECTS OF POLLUTANTS

In association with measurement and prediction of pollutant concentration, their health impact on human being also needs to be assessed. Out of various pollutants in atmosphere, carbon monoxides, particulate matter, nitrogen dioxide, sulfur dioxide, volatile organic compound, ozone and lead (Pb) are found as key air pollutants having adverse health effects. A good correlation between exposure to PM and cardiovascular deaths, myocardial infarctions and ventricular fibrillation has been found (Dockery,

Figure 8. Flow diagram of AERMOD modeling system (Source: Cimorelli, 2004)

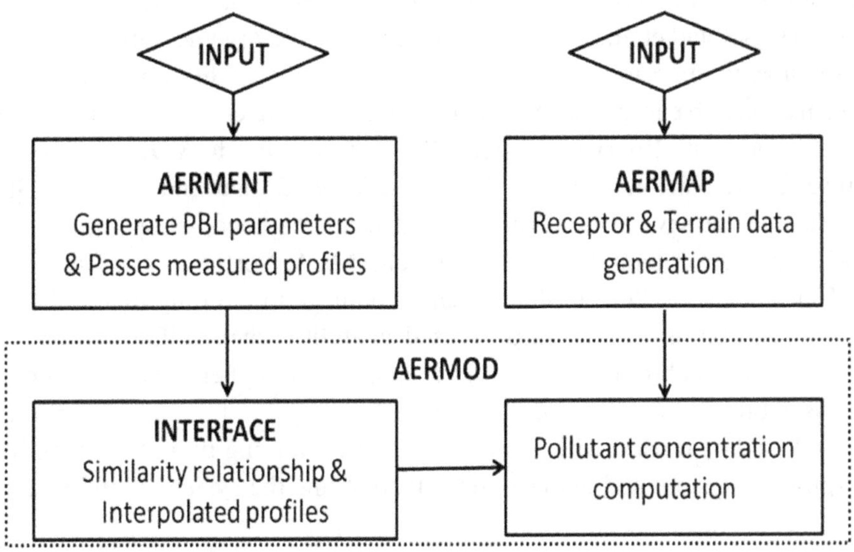

2001). (Mazumdar and Susman, 1983) reported the relation between the mortality by heart disease and PM exposure. Similarly, (Buckeridge et al., 2002) reveal the significant effect of $PM_{2.5}$ and PM_{10} on hospital admission rates for respiratory diagnoses. Respiratory diagnoses include bronchitis, chronic obstructive pulmonary disease, pneumonia, upper respiratory and lower respiratory tract infections. (Hsu C., et al. 2017) conducted a study for the measurement of ambient $PM_{2.5}$ in different seasons in residential area of central Taiwan. Study observed an increase of 10 µg/m^3 of $PM_{2.5}$ per day increases the respiratory physician visits by the 2%.

Exposure to CO can also have the significant health impact. Since the affinity of CO is 220 more than that of O_2, it does combine with hemoglobin and form the carboxyhemoglobin (COHb) (Haab, 1990). So formed COHb not only reduces the oxygen carrying capacity but also reduce the oxygen delivery at the tissue level. Individuals with cardiovascular disease are more susceptible to exposure to outdoor and indoor levels of CO. People with cardiovascular disease when subjected to the CO for 2- 4 hrs at the concentration around 50 ppm may feel angina pain post exercise and increase arrhythmias frequency (Aronow et al., 1972; Anderson et al., 1973; Allred et al., 1989). If the individual is exposed for the longer duration than it may result into the death. (Modic, 2003) reported the death of subject if exposed to 800 ppm for about 30 minutes. Exposures to CO (sufficient to reach blood COHb concentration of 4-6%) decrease exercise performance in young nonsmoking healthy individuals (Adir et al., 1999).

Similar, NO_2 also have the health effect in terms of airway infection and impair lung function. (Frampton et al., 2002) reveal the significant health impact of NO_2 and damage to the respiratory cells. A meta-analysis found that a 16 ppb increase in indoor NO_2 levels was associated with a 20% increased risk of respiratory illness in children (Hasselblad et al., 1992). Exposure to volatile organic compound (VOC), even for short time period, may lead to the eye, nose and throat irritation. VOC's long term exposure causes the cancer risk, kidney and liver damage, and sometime damage to the central nervous system. Sign of exposure to VOCs are headache, allergic effect, decline in serum cholinesterase levels, fatigue and dizziness. Studies have indicated that several VOCs are emitted during photocopier operation, causing indoor air quality problems (Hetes, R., Moore, M., & Northeim, C., 1995; Brown, 1999; Henschel et al., 2001). SO_2 can cause damage to building materials, statue, cloth, metals, and also result in decreased visibility when sulfate particles form in the air. Humans chronically exposed to SO_2 have a higher incidence of coughs, shortness of breath, bronchitis, fatigue, 'colds' of lung duration. Small particles can adsorb sulfur dioxide and with water form acid containing particles which irritate the respiratory system and damage the cells that line the system. Combined effect of SO_2 with small particles is worse than their individual effect on human health. Another pollutant Ozone (O_3) has become a significant pollutant as a result of increased population growth, industrial activities and automobile sector. It primarily targets the respiratory tract and results into lung function reduction and asthma. Consequently, respiratory ailment caused hospital admission to be increased.

Given the fact of adverse effect of lead (Pb) exposure on the nervous system and renal system a study was conducted in Taiwan by (Wang J., Karmaus J.J., and Yang C.C., 2017). This study observed the concentration of blood lead level and immunoglobuline E (IgE) in total 930 children with and without allergic diseases, and found positive correlation between Pb exposure and asthma in children. Also observed that the IgE level in blood facilitate the asthma by Pb.

FUTURE RESEARCH DIRECTION

Because of significant health impact of pollutant and global climate change, management of air quality is now days becoming a growing concern. CMAQ, AIRMOD and many more dispersion modeling software's can determine air quality of the particular region or state. IAQ can be predicted well using suitable modeling approach and can be researched for the new sources of pollutant inside the microenvironments. Furthermore, with the know condition of air quality associated with any climate situation and emission, suitable measure into the different sectors can be suggested to improve the air quality and reduce the pollution level. Also a comparative study can be conducted with the cities of different pollution level which gives a better insight of possible changes in the policies and program to improve the air quality of the concern area. .

CONCLUSION

Descriptive definition of exposure in association with target and agent is much more important in context of pollution studies. Discussion about the direct and indirect method of exposure measurement suggest that direct method of measurement can give better insight of the human exposure than the indirect method since it measure the pollution level to each individual. Better prediction of pollutants can be carried out using respective pollution models like for indoor, indoor model and for outdoor, ambient dispersion model. Since the prediction of pollutant concentration using different approach is gaining the attention of researchers, detail discussion about the fundamental components of two air quality models CAMQ and AIRMOD is reviewed for their capacity of predicting and analyzing the pollutant concentration.

A multiple scale modeling for different pollutant has made possible using CMAQ. It can incorporate the emission and metrological data from other sources as well and provide the facility of sensitivity analysis. AIRMOD is the steady state model based upon the Gaussian dispersion model and two preprocesses for meteorology and emission geographical data.

ACKNOWLEDGMENT

We sincerely thank to Ministry of Human Resource Development, Government of India for providing research assistantship which encourage writing this manuscript. We would also like to thank the reviewers for their constructive suggestions to improve this manuscript.

REFERENCES

Adir, Y., Merdler, A., Haim, S. B., Harduf, R., & Bitterman, H. (1999). Effects of exposure to low concentrations of carbon monoxide on exercise performance and myocardial perfusion in young healthy men. *Occupational and Environmental Medicine*, *56*(8), 535–538. doi:10.1136/oem.56.8.535 PMID:10492650

Akland, G. G., Ott, W. R., & Wallace, L. A. (1984). *Human exposure assessment: Background concepts, purpose, and overview of the Washington, DC.-Denver, Colorado Field Studies. NTIS.*

Allred, E. N., Bleecker, E. R., Chaitman, B. R., Dahms, T. E., Gottlieb, S. O., Hackney, J. D., & Warren, J. et al. (1989). Short-term effects of carbon monoxide exposure on the exercise performance of subjects with coronary artery disease. *The New England Journal of Medicine*, *321*(21), 1426–1432. doi:10.1056/NEJM198911233212102 PMID:2682242

Anderson, E. W., Andelman, R. J., Strauch, J. M., Fortuin, N. J., & Knelson, J. H. (1973). Effect of low-level carbon monoxide exposure on onset and duration of angina pectorisa study in ten patients with ischemic heart disease. *Annals of Internal Medicine*, *79*(1), 46–50. doi:10.7326/0003-4819-79-1-46 PMID:4578639

Aronow, W. S., Harris, C. N., Isbell, M. W., Rokaw, S. N., & Imparato, B. (1972). Effect of freeway travel on angina pectoris. *Annals of Internal Medicine*, *77*(5), 669–676. doi:10.7326/0003-4819-77-5-669 PMID:4117097

Brown, S. K. (1999). Assessment of Pollutant Emissions from Dry-Process Photocopiers. *Indoor Air*, *9*(4), 259–267. doi:10.1111/j.1600-0668.1999.00005.x PMID:10649859

Buckeridge, D. L., Glazier, R., Harvey, B. J., Escobar, M., Amrhein, C., & Frank, J. (2002). Effect of motor vehicle emissions on respiratory health in an urban area. *Environmental Health Perspectives*, *110*(3), 293–300. doi:10.1289/ehp.02110293 PMID:11882481

Byun, D., & Schere, K. L. (2006). Review of the governing equations, computational algorithms, and other components of the models-3 community multiscale air quality (CMAQ) modeling system. *Applied Mechanics Reviews*, *59*(2), 51–77. doi:10.1115/1.2128636

Cernichiari, E., Brewer, R., Myers, G. J., Marsh, D. O., Lapham, L. W., Cox, C., & Clarkson, T. W. et al. (1994). Monitoring methylmercury during pregnancy: Maternal hair predicts fetal brain exposure. *Neurotoxicology*, *16*(4), 705–710. PMID:8714874

Ching, J., & Byun, D. (1993). *Introduction to the models-3 framework and the community multiscale air quality model (CMAQ)*. EPA/600/R-99/030.

Cimorelli, A.J. (2004). *AERMOD: Description of model formulation*. EPA-454/R-03-004.

Clayton, C. A., Perritt, R. L., Pellizzari, E. D., Thomas, K. W., Whitmore, R. W., Wallace, L. A., & Spengler, J. D. (1992). Particle Total Exposure Assessment Methodology (PTEAM) study: Distributions of aerosol and elemental concentrations in personal, indoor, and outdoor air samples in a southern California community. *Journal of Exposure Analysis and Environmental Epidemiology*, *3*(2), 227–250. PMID:7694700

Devi, J. J., Gupta, T., Jat, R., & Tripathi, S. N. (2013). Measurement of personal and integrated exposure to particulate matter and co-pollutant gases. *Environmental Science and Pollution Research International*, *20*(3), 1632–1648. doi:10.1007/s11356-012-1179-3 PMID:22965544

Dockery, D., Pope, A., Xu, X., Spengler, J., Ware, J., Fay, M., & Speizer, F. et al. (1993). An association between air pollution and mortality in six U.S. cities. *The New England Journal of Medicine*, *329*(24), 1753–1759. doi:10.1056/NEJM199312093292401 PMID:8179653

Dockery, D. W. (2001). Epidemiologic evidence of cardiovascular effects of particulate air pollution. *Environmental Health Perspectives*, *109*(Suppl 4), 483–486. doi:10.1289/ehp.01109s4483 PMID:11544151

Duan, N. (1982). Models for human exposure to air pollution. *Environment International*, *8*(1-6), 305–309. doi:10.1016/0160-4120(82)90041-1

Environmental Protection Agency. (1992). Guidelines for exposure assessment. *Federal Register*, *57*(104), 22887–22938.

Frampton, M. W., Boscia, J., Roberts, N. J., Azadniv, M., Torres, A., Cox, C., & Gibb, F. R. et al. (2002). Nitrogen dioxide exposure: Effects on airway and blood cells. *American Journal of Physiology. Lung Cellular and Molecular Physiology*, *282*(1), L155–L165. PMID:11741827

Goyal, R., & Khare, M. (2010). 4 Indoor Air Pollution and Health Effects. *Health and Environmental Impacts*, 109.

Haab, P. (1990). The effect of carbon monoxide on respiration. *Cellular and Molecular Life Sciences*, *46*(11), 1202–1206. doi:10.1007/BF01936937 PMID:2174793

Hasselblad, V., Eddy, D. M., & Kotchmar, D. J. (1992). Synthesis of environmental evidence: Nitrogen dioxide epidemiology studies. *Journal of the Air & Waste Management Association, 42*(5), 662–671. doi:10.1080/10473289.1992.10467018 PMID:1627322

Henschel, D. B., Fortmann, R. C., Roache, N. F., & Liu, X. (2001). Variations in the emissions of volatile organic compounds from the toner for a specific photocopier. *Journal of the Air & Waste Management Association, 51*(5), 708–717. doi:10.1080/10473289.2001.10464309 PMID:11355458

Hetes, R., Moore, M., & Northeim, C. (1995). Office equipment: design, indoor air emissions, and pollution prevention opportunities. In *Office equipment: design, indoor air emissions, and pollution prevention opportunities*. EPA.

Hsu, C. Y., Chiang, H. C., Chen, M. J., Chuang, C. Y., Tsen, C. M., Fang, G. C., & Chen, Y. C. et al. (2017). Ambient PM 2.5 in the residential area near industrial complexes: Spatiotemporal variation, source apportionment, and health impact. *The Science of the Total Environment, 590*, 204–214. doi:10.1016/j.scitotenv.2017.02.212 PMID:28279531

IEHIAS. (n.d.). Retrieved from http://www.integratedassessment.eu/eu/guidebook/indoor_air_pollution_models.html

Krotoszynski, B. K., Bruneau, G. M., & O'Neill, H. J. (1979). Measurement of chemical inhalation exposure in urban population in the presence of endogenous effluents. *Journal of Analytical Toxicology, 3*(6), 225–234. doi:10.1093/jat/3.6.225

Lioy, P. J. (1990). Assessing total human exposure to contaminants. *Environmental Science & Technology, 24*(7), 938–945. doi:10.1021/es00077a001

Louie, A. H., & Pierce, R. C. (1988). Mathematical Models of Human Exposure to air pollutants. *Mathematical and Computer Modelling, 10*(1), 49–64. doi:10.1016/0895-7177(88)90121-5

Mandal, B. K., Ogra, Y., & Suzuki, K. T. (2003). Speciation of arsenic in human nail and hair from arsenic-affected area by HPLC-inductively coupled argon plasma mass spectrometry. *Toxicology and Applied Pharmacology, 189*(2), 73–83. doi:10.1016/S0041-008X(03)00088-7 PMID:12781625

Mazumdar, S., & Sussman, N. (1983). Relationships of air pollution to health: Results from the Pittsburgh study. *Archives of Environmental Health: An International Journal, 38*(1), 17–24. doi:10.1080/00039896.1983.10543974 PMID:6830314

Modic, J. (2003). Carbon monoxide and COHb concentration in blood in various circumstances. *Energy and Building, 35*(9), 903–907. doi:10.1016/S0378-7788(03)00022-7

Monn, C. (2001). Exposure assessment of air pollutants: A review on spatial heterogeneity and indoor/outdoor/personal exposure to suspended particulate matter, nitrogen dioxide and ozone. *Atmospheric Environment, 35*(1), 1–32. doi:10.1016/S1352-2310(00)00330-7

Nazaroff, W. W. (2004). Indoor particle dynamics. *Indoor Air, 14*(s7), 175–183. doi:10.1111/j.1600-0668.2004.00286.x PMID:15330785

Ott, W. R., Steinemann, A. C., & Wallace, L. A. (Eds.). (2006). *Exposure analysis*. CRC Press. doi:10.1201/9781420012637.pt1

Pepper, D. W., & Carrington, D. (2009). *Modeling indoor air pollution*. Imperial College Press. doi:10.1142/p612

Pitarma, R. A., Ramos, J. E., Ferreira, M. E., & Carvalho, M. G. (2004). Computational fluid dynamics: An advanced active tool in environmental management and education. *Management of Environmental Quality, 15*(2), 102–110. doi:10.1108/14777830410523053

Pollutants, I. (1981). *Committee on Indoor Pollutants*. National Research Council.

Spengler, J. D., Samet, J. M., & McCarthy, J. F. (2001). *Indoor air quality handbook*. New York: McGraw-Hill.

Spengler, J. D., & Sexton, K. (1983). Indoor air pollution—a public health perspective. *Science, 221*(4605), 9–17. doi:10.1126/science.6857273 PMID:6857273

Vedal, S. (1997). Critical review-Ambient particles and health: Lines that divide. *Journal of the Air & Waste Management Association, 47*(5), 551–581. doi:10.1080/10473289.1997.10463922 PMID:9155246

Wang, C., Cai, J., Chen, R., Shi, J., Yang, C., Li, H., & Xia, Y. et al. (2017). Personal exposure to fine particulate matter, lung function and serum club cell secretory protein (Clara). *Environmental Pollution, 225*, 450–455. doi:10.1016/j.envpol.2017.02.068 PMID:28284549

Wang, I. J., Karmaus, W. J., & Yang, C. C. (2017). Lead exposure, IgE, and the risk of asthma in children. *Journal of Exposure Science & Environmental Epidemiology, 27*(5), 478–483. doi:10.1038/jes.2017.5 PMID:28401896

Wayne, R. O. (1982). Concepts of human exposure to air pollution. *Environment International*, *7*(3), 179–196. doi:10.1016/0160-4120(82)90104-0

Yanagisawa, Y., Nishimura, H., Matsuki, H., Osaka, F., & Kasuga, H. (1988). Urinary hydroxyproline to creatinine ratio as a biological effect marker for exposure to NO2 and tobacco smoke. *Atmospheric Environment, 22*(10), 2195-2203.

Zierler, K. (1981). A critique of compartmental analysis. *Annual Review of Biophysics and Bioengineering*, *10*(1), 531–562. doi:10.1146/annurev.bb.10.060181.002531 PMID:7259129

KEY TERMS AND DEFINITIONS

Control Volume: It is the imaginary boundary that isolates the part of the system and then over the surface of the boundary conservation equations are applied.

Emission Inventory: It keeps the information of the amount of different pollutants discharged into the atmosphere.

Meteorological Modeling: It is the mathematical formulation of various meteorological processes.

Planetary Boundary Layer: It is the lowest part of the troposphere and directly influenced by the various earth forcings.

$PM_{1.0}$: Particulate matter less than or equal to 1 μm size.

Sensitivity Analysis: It is the measure of the degree of variability of the output of the system with respect to change in the input variables.

Chapter 7
Using of Modified SBA-15 Mesoporous Silica Materials for CO2 Capture:
A Review

Filiz Akti
Hittite University, Turkey

ABSTRACT

Carbon dioxide emissions cause global warming, and greenhouse gases and climate change are very serious problems. Mesoporous silica material SBA-15 has been preferred mostly as an ideal adsorbent for CO2 due to its excellent properties such as high surface areas and pore volumes, larger pore diameter, and thicker silica wall. In the literature studies, SBA-15 has been modified by different functional groups and the effects of modification methods on the CO2 adsorption have been investigated. Modified SBA-15 adsorbents showed high CO2 adsorption capacity. The aim of this chapter is to review the use of modified-SBA-15 mesoporous silica materials as adsorbent for CO2 capture.

INTRODUCTION

Carbon dioxide emission from fossil fuel combustion is one of the major causes of global warming and climate change and also has to be decreased in order to stabilize the CO_2 concentration level in the atmosphere. Therefore, many researchers focus on searching of efficient and economical methods to control CO_2 emissions. Various physical and chemical methods, such as adsorption, solvent absorption, membrane

DOI: 10.4018/978-1-5225-3379-5.ch007

separation, and cryogenic distillation, have been used for CO_2 capture (Zhang et al., 2011, Sanz et al., 2012, Zhang et al., 2013, Lakhi et al., 2016, Shi-Yaumi et al., 2017). The traditionally and commercially available technique for CO_2 capture is chemical absorption with liquid alkanolamines, such as monoethanolamine, diethanolamine, and methyldiethanolamine (Gil et al., 2011, Sanz et al., (2012). The use of liquid amines includes many disadvantages, such as corrosion of processing equipment and high energy consumption due to solvent regeneration (Sanz et al., 2012, Yan et al., 2013, Sanz-Pérez et al., 2013). The adsorption method is preferred potentially due to it having a simple, energy-efficient, and inexpensive separation process. Solid adsorbents, having lower heat capacity, reduce the amount of necessary energy for generation as compared to aqueous amine solutions. On the other hand, capacity losses upon cycling and corrosion problems are eliminated with the use of solid adsorbents (Yan et al., 2013, Liu et al., 2017). Zeolite, activated carbon, and silica mesoporous materials, which have large and uniform pores, tunable pore sizes, and large surface areas, are given much attention for the CO_2 capture. Zeolites, having microporous structure, large surface areas, and specific sites of adsorption, are mostly preferred in the gas separation adsorbents at low temperature. But gas diffusion to micropores of zeolites is difficult due to mass transfer limitation Linfang et al., 2007, Zhou et al., 2013, Chen et al., 2014, Shakerian et al., 2015). In recent years, ordered mesoporous silica materials, such as MCM-41, MCM-48, and SBA-15, have attracted attention for the CO_2 capture studies due to their having high surface areas, uniform and tunable large pore sizes, and surface functional groups (Wei et al., 2008, Yan et al., 2013, Yaumi et al., 2017). These mesoporous silica materials are used generally by modified with basic amine groups in the CO_2 capture. The modification of silica materials are performed by chemical (grafting and co-condensation method) or physical (impregnation method) surface modification methods. (Sanz et al., 2012, Yan et al., 2013, Ullah et al., 2015). The SBA-15 material as compared with other materials includes micropores together with mesopores in the silica wall and these micropores provide high surface area and high gas separation selectivity (Zhang et al., 2013, Yan et al., 2013). In the CO_2 capture studies performed with SBA-15, modification and functionalization treatments applied to the SBA-15 significantly increase CO_2 sorption due to the formation of carbamates. The modifications and functionalize treatments have been provided by amine groups (Da'na & Sayari, 2011, Sanz et al., 2012, Yan et al., 2013, Sanz-Pérez et al., 2013, Ullah et al., 2015, Sanz-Pérez et al., 2016), polyethylenimine (Wang & Song, 2012), aminopropyltriethoxysilane (Wang et al., 2007, Wei et al., 2010), octa(3-aminophenyl)octasilsesquioxane, 3-chloropropyltrimethoxysiliane (CPTMS) (Bhagiyalakshmi et al., 2010), amine dendrimers (Jing et al., 2014), carboxylic acid (Khatun et al., 2017), and zirconia (Thunyaratchatanon et al., 2017).

The main objective of the present study is to investigate the CO_2 capture performance of the amine-modified SBA-15 material and the effects of amine type, modification methods, and the CO_2 adsorption mechanism.

SBA-15 MATERIALS AS CO_2 CAPTURE SORBENTS

Properties and Synthesis Method of SBA-15

SBA-15 (Santa Barbara Amorphous No: 15) is mesoporous material having highly ordered uniform hexagonal mesoporous structures, high surface areas (600- 1000 m^2/g), large pore size diameters (4.6–30 nm), and thicker silica wall (3-9 nm). In addition, it exhibits thermal and hydrothermal stability due to thicker silica wall thickness. The presence of micropores in the walls between the mesopores provides advantages such as higher surface areas and reduction of the diffusion limitation. In the synthesis of SBA-15, non-ionic triblock copolymer as a template, is used and also tetraethylorthosilicate (TEOS), tetramethylorthosilicate (TMOS), or tetrapropylorthosilicate (TPOS) as a silica source. The SBA-15 materials synthesize under acidic conditions, and strong acid media (< pH=1) provides by different acids source, such as HCl, HBr, HI, HNO_3, H_2SO_4, or H_3PO_4 (Zhao et al., 1998, Øye et al., 2001, Shah et al., 2007, Zhao et al., 2009, Huirache-Acuña et al., 2013). For a typical synthesis (Zhao et al., 1998), Pluronic P123 (polyethylene oxide (PEO)–polypropylene oxide (PPO)–polyethylene oxide (PEO)) is a good candidate to create a template for the silica wall due to it having ordered mesostructure properties, amphiphilic character, and low cost. The template is dissolved in a solution of water and HCl, and then the required amount of TEOS is added between temperatures of 35-40 °C. Thus, the polymerization and condensation steps occur by the formation of silanol groups with the hydrolysis of the alkoxide precursors in low-pH, where the interaction occurs through the $S^oH^+X^-I^+$ mechanisms given below:

1. Polymerization step; the formation of silanol groups by hydrolysis of the alkoxide precursors
   $\equiv Si - OR + H_2O \leftrightarrow Si - OH + ROH$
2. Condensation step; the formation of water or alcohol
   $\equiv Si - OH + HO\ Si \equiv \leftrightarrow \equiv Si - O - Si \equiv + H_2O$
   $\equiv Si - OH + RO - Si \equiv \leftrightarrow \equiv Si - O - Si + ROH$

where S^oH^+ is the surfactant hydrogen bonded to a hydronium ion, X^- the chloride ion, and I^+ the protonated silica. An aqueous solution of triblock copolymer and TEOS is kept under stirring conditions for 24 h for aging. The mixture is subsequently aged at 100 °C for 48 h in an autoclave. Finally, the obtained solid is washed with deionized water; filtered, dried in air at room temperature and then calcined. The calcination treatment is carried out by slowly increasing temperature from room temperature to 550 °C in 6 h. The latter step of the template removal is one of the crucial aspects in the synthesis of ordered mesoporous, because the procedure employed during calcination influences the final textural properties of SBA-15 material. The synthesis mechanism of SBA-15 is given in Figure 1.

Modification and Functionalize Treatments of SBA-15

The modification of silica mesoporous materials such as SBA-15 used for high CO_2 capture performance is usually performed by incorporating functional groups with high affinity. The modification treatment of SBA-15 is especially performed with amino groups. Thus, high effective interactions occur between weakly acidic CO_2 with the basic amino groups, which is incorporated into the SBA-15 structure (Chen et al., 2014). A modification treatment is mostly performed by chemical or physical methods. The chemical modification is performed with the amino groups (Sanz et al., 2012, Zhang et al., 2013), such as propylamine (Jing et al., 2014), hydroxylamine (Ulllah et al., (2015), aminopropyl (Sanz et al., 2012, Khatun et al., 2017), ethylene-diamine propyl (Sanz et al., 2012), diethylene-triaminepropy (Sanz et al.,2012), polyethyleneimine (Sanz et al., 2012, Yan et al., 2013), and tetraethylenepentamine (Zhang et al., 2013). A high concentration of hydroxyl groups on the mesoporous silica surface is favorable for achieving a high amine species loading (Chen et al., 2014). Thus, the present study especially focuses on modification studies performed with the amino group.

The modification treatments are performed with the use of different routes, such as impregnation, post-synthesis grafting, and direct co-condensation methods. In the

Figure 1. Synthesis mechanism of SBA-15 (Myers et al., 1992)

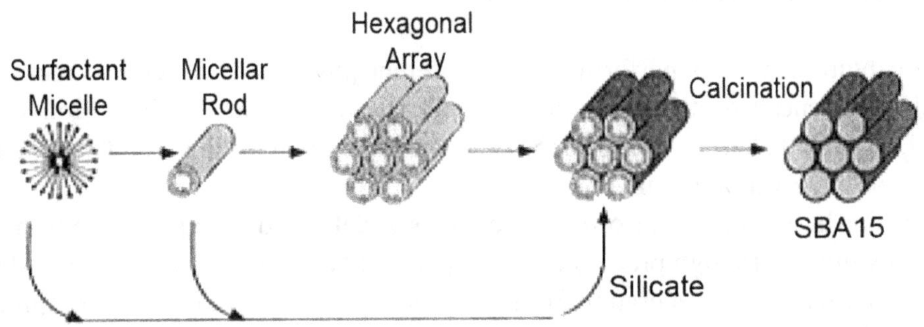

impregnation method, solid support materials are mixed with amino groups. The amino group source is dissolved using a polar solvent so that the amine species can be better incorporated into the channel of the support, and then the excess solvent is evaporated. Impregnation is a simple and mild synthesis method, but amine groups are only physically settled into the pore structure of silica support and this might prevent the stability of the amine-support materials. On the other hand, the impregnation method provides a high amount of amine incorporated into the support due to the high pore volume and surface area of silica support materials. The grafting method is based on the reaction between the silanol groups on the silica surface and amino groups which are covalently bonded to the silanol groups and immobilized onto silica support. The grafting method provides more thermal stability as compared the impregnation method. In the grafting synthesis of amino-silica support, silica is dissolved into a solvent such as toluene, mixed with amino groups and the obtained mixture is heated under reflux. The resulted solid product is washed to remove the unreacted amino groups. In adsorbents that are synthesized with this method, the amount of incorporated amine is related to the number of hydroxyl groups on the silica surface. Direct co-condensation or direct hydrothermal method involves the simultaneous condensation of the organosilanes and aminosilanes precursor in the synthesis medium in the presence of an organic template and acid or base catalyst (Sanz et al., 2012, Chen et al., 2014, Unveren, et al., 2017, Chen et al., 2017).

The advantages of impregnation over the grafting method are mainly its practical preparation, high amine capacity, and decreased corrosion behaviour (Sanz et al., 2012, Zhang et al., 2013, Samiey et al., 2014, Unveren, et al., 2017).

Textural properties of SBA-15 and SBA-15 modified with different functional groups are given in Table 1. It is seen that impregnation and grafting methods are generally used for modification of SBA-15 with amino groups, and these modification treatments caused a decrease in surface area, pore volume, and pore diameter of SBA-15 due to pore blockage (Yaumi et al., 2017).

CO_2 CAPTURE AND MECHANISM

The capture of CO_2 is performed with different physical and chemical methods, such as solvent absorption, adsorption, membrane separation, and distillation (Chen et al., 2014, Shakerian et al., 2015). The mostly commonly applied CO_2 capture methods are given in Figure 2.

Amine based solvent absorption process is widely used in CO_2 capture from the gas mixture due to high process efficiency. But it has disadvantages, such as high energy demand for regeneration of the solvent, corrosion of processing equipment, and high volatility. Therefore adsorption processes using solid sorbents is more

Table 1. Textural properties of SBA-15 and SBA-15 modified with different functional groups

Modification method	Functional groups	Textural properties			Reference
		BET surface area (m^2/g)	Pore diameter (nm)	Pore volume (cm^3/g)	
	–	775	8.9	1.10	Sanz et al., 2012
Grafting	Diethylenetriaminopropyl	166	5.8	0.30	
Impregnation	Polyethyleneimine (PEI)	49	5.9	0.09	
	–	711	7.7	1.11	Ullah et al., 2015
Impregnation	NH_2OH	433	6.8	0.54	
	–	640	5.6	0.9	Zhang et al., 2013
Impregnation	Tetraethylenepentamine (TEPA)	0.86	2.3	0.002	
	–	782	9.8	0.91	Zhang et al., 2011
Grafting	Zirconium phosphate-5′-adenylic acid	478	6.5	0.57	
	–	894	6.2	1.12	
Direct hydrothermal	Zirconium	729	6.1	0.81	Thunyaratchatanon et al., 2017
	–	587	8.5	0.96	Sanz-Perez et al., 2013
Impregnation	Tetraethylenepentamine (TEPA)	186	7.0	0.34	
	–	582	9.2	0.95	Linfang et al., 2007
Grafting	3-aminopropyltriethoxysilane (APTES)	313	7.4	0.57	
	–	752	3.7	0.72	Bhagiyalakshmi et al., 2010
Grafting	Octa (aminophenyl) silsesquioxane	86.2	2.0	0.13	
	–	700	6.1	1.00	Chandrasekar et al., 2009
Wet impregnation	Polyethyleneimine (PEI)	407	7.2	0.70	
	–	802	8.6	1.14	Yan et al., 2011
Wet impregnation	Polyethyleneimine (PEI)	46	4.9	0.11	

often preferred. Adsorption is a well-known separation method and recognized as one of most efficient and economic methods. In addition, owing to the reversible nature of most adsorption processes, the adsorbents can be regenerated by suitable desorption processes for multiple uses, and many desorption processes have low maintenance cost, high efficiency, and easy operation. But the major problem in this

Figure 2. CO_2 capture methods (Yaumi et al., 2017)

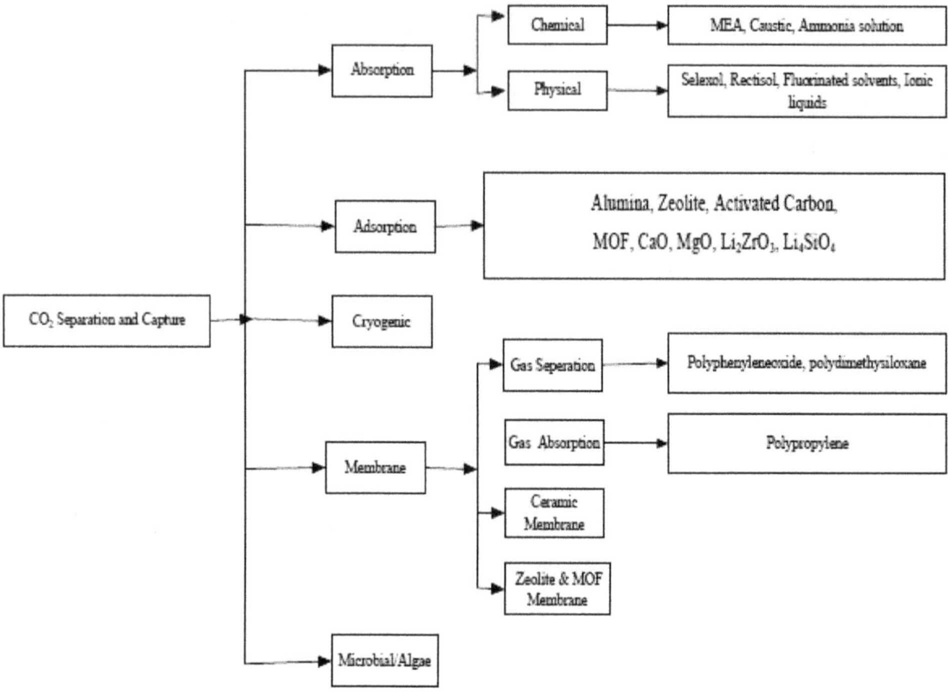

field is to select novel types of adsorbents having large specific surface areas, high selectivity, and high regeneration ability (Leung et al., 2014). The high capture of CO_2, which is a weak Lewis acid, largely depends on surface chemistry and physical properties, such as surface functional groups, surface area, and porous structure of the materials. The surface functionalities are achieved via impregnation using chemicals such as NH_3, HNO_3, and amines or introduction of nitrogen precursors onto support. (Yaumi et al., 2017).

The SBA-15 sorbents treated with different modification methods and functional groups investigated for CO_2 capture and results are given in Table 2.

Generally, the modification of SBA-15 is carried out with impregnation and grafting methods. The modification with amino groups caused a decrease of textural properties of SBA-15. The textural mesoporosity in mesoporous silica supports is important for CO_2 capture. This leads to an increase in the incorporation potential of amine species to the pore structure and facilitates CO_2 diffusion inside the pore channels. Especially PEI and TEPA modification is effect positively the capacity of CO_2 capture. The CO_2 adsorption capacity of PEI and TEPA impregnated SBA-15 sorbents, which have a lower surface area and pore diameter, and higher pore volume

Table 2. CO_2 capture capacity of SBA-15 support material modified with different functional groups

Modification method	Functional groups	CO_2 adsorption		CO_2 adsorption capacity	Reference
		Parameters	Technique		
Grafting	Diethylenetriaminopropyl	0.15 atm, 45°C	Adsorption/ desorption isotherms	60.8 mg/g	Sanz et al., 2012
Impregnation	Polyethyleneimine (PEI)	0.15 atm, 45°C	Adsorption/ desorption isotherms	71.2 mg/g	
Impregnation	NH_2OH	1 bar, 25°C	High pressure	72.6 mg/g	Ullah et al., 2015
Impregnation	Tetraethylenepentamine (TEPA)	Atmospheric pressure, 25°C	Flow adsorber	171.8 mg/g	Zhang et al., 2013
Grafting	Zirconium phosphate-5′-adenylic acid	1 atm, 25°C	DSC-TGA	87.6 mg/g	Zhang et al., 2011
Direct hydrothermal	Zirconium	1 bar, 30°C	Volumetric sorption	29.92 mg/g	Thunyaratchatanon et al., 2017
Impregnation	Tetraethylenepentamine (TEPA)	1 bar, 25°C	TGA	9.77 mg/ mg	Sanz-Perez et al., 2013
Wet impregnation	Polyethyleneimine (PEI)	1 bar, 75°C,	TPD	194 mg/g	Wang & Song, 2012
Grafting	3-aminopropyltriethoxysilane (APTES)	0.005 MPa, room temperature	microelectronic recording balance system	34.2 mg/g	Linfang et al., 2007
Grafting	Octa (aminophenyl) silsesquioxane	Atmospheric pressure, 25°C	TGA	83 mg/g	Bhagiyalakshmi et al., 2010
Wet impregnation	Polyethyleneimine (PEI)	Atmospheric pressure, 75°C	TG	120 mg/g	Chandrasekar et al., 2009
Wet impregnation	Polyethyleneimine (PEI)	Atmospheric pressure, 75°C	Fixed bed reactor	105.2 mg/g	Yan et al., 2011
Grafting	Diamine	Atmospheric pressure, 120 °C	MS and TPD	34.6 mg/g	Khatri et al., 2005

than that of SBA-15 sorbents modified with the other amino groups (Table 1 and 2). Probably this behavior is due to the high N atom concentration of PEI and TEPA (PEI and TEPA have N contents of ca. 33% and 37%, respectively). PEI can be found in several forms, linear and branched, with different molecular weights. Loading different types of PEI into porous silica affects the CO2 capture performance of amine–silica composites (Wang & Song, 2012, Sanz et al., 2012, Yan et al., 2013, Chen et al., 2014, Chen et al., 2017). The high capacity of PEI-based sorbents can be attributed to increasing the interaction between PEI-modified support and CO_2 due to high surface area and a large number of active sites (Wang & Song, 2014). In the study of Ullah et al., it was stated that amine modification caused a decrease in surface areas and pore volume values of sorbents, but CO_2 adsorption was obtained as high due to the high affinity of NH_2 molecules at low pressure and temperature. On the other hand, the CO_2 adsorption capacity increased with the increase of the microporosity in the pure SBA-15 sorbents, but the role of microporoes is not clear due to the incorporation of PEI in the pores as they may have been blocked by amine. In this case, the CO_2 uptake of amine-modified SBA-15 sorbents with and without micropores correlated with their surface areas (Yan et al., 2013). Zhang et al., (2013) stated that as the amount of amine increased CO_2, adsorption increased, but CO_2 adsorption decreased above 70 wt % amine loading. When the amine loading exceeded the pore saturation level, the adsorption process is controlled by diffusion and this caused a decrease in CO_2 sorption. On the other hand if the organic amines do not completely fill up pore channels and they are stayed on the external surface of particles, the interaction of CO_2 molecules with active sites of amines prevent which caused a decrease in adsorption capacity. The mechanism model proposed by Wang & Song, (2012) in relation to this issue is given in Figure 3. In this model, layers which are named as the exposed PEI layer and the bulky PEI layer vary with the change of PEI loading and sorption temperature. At 75 °C, when PEI loading is lower (probably less than 44 wt %), only the exposed PEI layer exists. If the PEI

Figure 3. Proposed CO_2 sorption model over PEI modified-SBA-15 sorbent (Wang & Song, 2012)

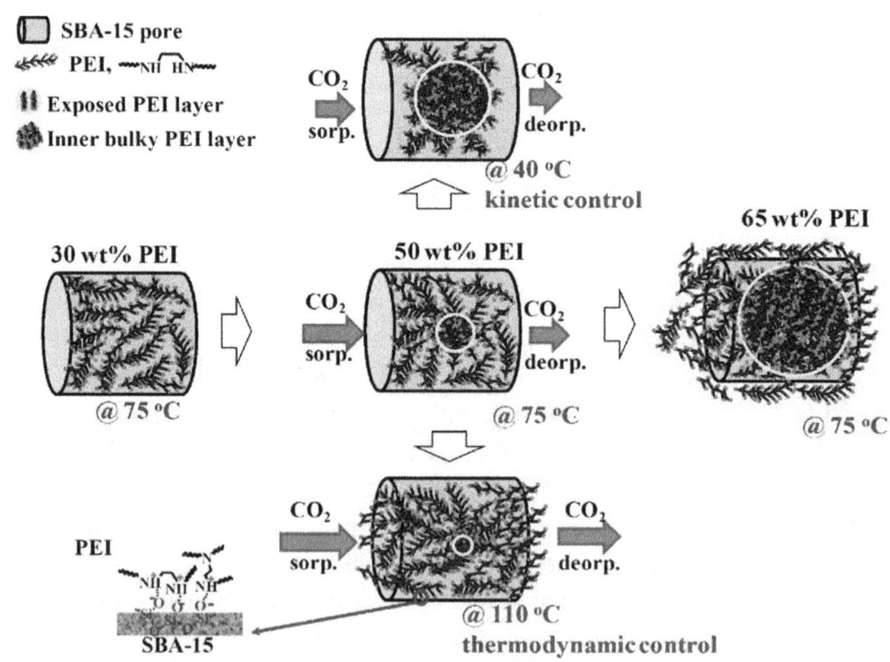

content is more than 50 wt %, the PEI loadings will be higher and the bulky PEI layer will be larger. When the sorption temperature changes, the loaded PEI polymer changes accordingly, stretching at high temperatures and shrinking at low temperatures as the polymer structure is flexible. As for CO_2 sorption over PEI/SBA-15 sorbent at 75 °C, the total amine sites increase with the increase of PEI loading. The CO_2 capture by amines is an exothermic process, and a decrease in temperature decreases adsorption capacity because adsorption is a thermodynamically controlled process.

Figure 4 shows the presence of two types of interactions occurring between amine groups and silica support. One of the interactions is between acidic CO_2 molecules and basic amine groups to form carbonate. The other, the reaction between amine, CO_2, and water led to bicarbonate formation that alters the surface chemistry and develops specific sites for CO_2 adsorption. (Unveren et al., 2017). On the other hand, water can influence CO_2 adsorption through both kinetic and thermodynamic factors. Water has been shown to both enhance and hurt CO_2 adsorption, depending

Figure 4. Reaction pathway of primary amine groups and CO_2 molecules (Sanz-Perez et. al, 2013)

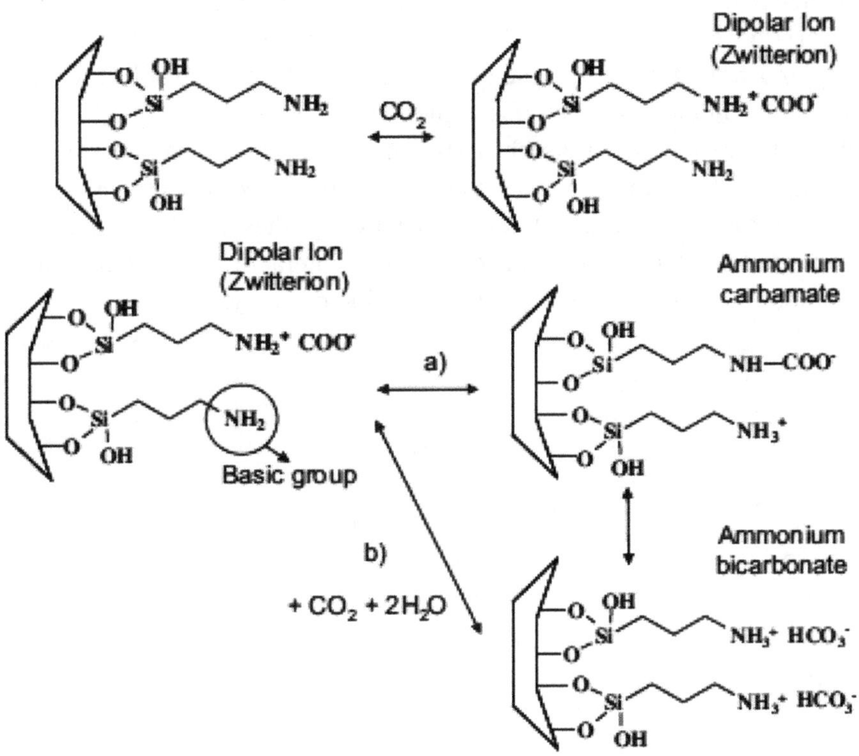

on the adsorbent and conditions used. Among kinetic factors, in conventional oxide-supported amine adsorbents water can alter the accessibility of amines by improving or impeding CO_2 transport through amine layers (Darunte et al., 2016). In the study carried out by the SBA-15 modified with 3-aminopropylsilyl groups, the effect of water on the amine efficiency for CO_2 capture has been investigated and it has been seen that the maximum efficiency increase at low amine loading (Didas et al., 2014). If silica-based CO_2 adsorbents are impregnated with amine, the affinity of adsorbents to water vapor in gas feed increases. Therefore, amine modified mesoporous silica is expected to be a promising adsorbent with high CO_2 adsorption performance even in wet conditions (Chen et al., 2017, Hori et al., 2017).

CONCLUSION

For the current situation of CO_2 capture, mesoporous silica adsorbents modified with different functional groups are at the forefront in the development of new sorbents that could eliminate both the economical and high energy requirements for regeneration. SBA-15 has been considered in the present study due to its excellent physicochemical properties. SBA-15 modified by basic amino groups showed high CO_2 adsorption capacity. The grafting and impregnation methods are preferred generally, and the CO_2 capture capacity of SBA-15 modified with polyethyleneimine (PEI) by the impregnation method is higher.

In the literature review, it was seen that the amine modification treatment performed mostly with single amine groups. Future research related to CO_2 capture should focus on the following areas: (1) development of synthesis methods that preserve textural properties (2) the formation of synergistic interaction with double and triple modification so as to increase the CO_2 adsorption capacity.

REFERENCES

Bhagiyalakshmi, M., Anuradha, R., Park, S. D., & Tae, H. J. (2010). Octa (aminophenyl) silsesquioxane fabrication on chlorofunctionalized mesoporous SBA-15 for CO_2 adsorption. *Microporous and Mesoporous Materials*, *131*(1-3), 265–273. doi:10.1016/j.micromeso.2010.01.001

Chen, C., Kim, J., & Ahn, W. S. (2014). CO_2 capture by amine-functionalized nanoporous materials: A review. *Korean Journal of Chemical Engineering*, *31*(11), 1919–1934. doi:10.1007/s11814-014-0257-2

Chen, C., Park, D. W., & Ahn, W. S. (2014). CO_2 capture using zeolite 13X prepared from bentonite. *Applied Surface Science*, *292*, 63–67. doi:10.1016/j.apsusc.2013.11.064

Chen, C., Zhang, S., Row, K. H., & Ahn, W.S. (2017). *Amine–silica composites for CO_2 capture: A short review*. Academic Press. 10.1016/j.jechem.2017.07.001

Da'na, E., & Sayari, A. (2011). Adsorption of copper on amine-functionalized SBA-15 prepared by co-condensation: Equilibrium properties. *Chemical Engineering Journal*, *166*(1), 445–453. doi:10.1016/j.cej.2010.11.016

Darunte, L. A., Walton, K. S., Sholl, D. S., & Jones, C. W. (2016). CO_2 capture via adsorption in amine-functionalized sorbents. *Current Opinion in Chemical Engineering*, *12*, 82–90. doi:10.1016/j.coche.2016.03.002

Desmond Ng, J. W., Zhong, Z., Luo, J., & Borgna, A. (2010). Enhancing preferential oxidation of CO in H_2 on Au/α-Fe_2O_3 catalyst via combination with APTES/SBA-15 CO_2-sorbent. *International Journal of Hydrogen Energy*, *35*(23), 12724–12732. doi:10.1016/j.ijhydene.2010.08.136

Didas, S. A., Zhu, R., Brunelli, N. A., Sholl, D. S., & Jones, C. W. (2014). Thermal, oxidative and CO_2 induced degradation of primary amines used for CO_2 capture: Effect of alkyl linker on stability. *The Journal of Physical Chemistry C*, *118*(23), 12302–12311. doi:10.1021/jp5025137

Gil, M., Tiscornia, I., Iglesia, Ó., Mallada, R., & Santamaría, J. (2011). Monoamine-grafted MCM-48: An efficient material for CO_2 removal at low partial pressures. *Chemical Engineering Journal*, *175*, 291–297. doi:10.1016/j.cej.2011.09.107

Heydari-Gorji, A., Yang, Y., & Sayari, A. (2011). Effect of the pore length on CO_2 adsorption over amine-modified mesoporous silicas. *Energy & Fuels*, *25*(9), 4206–4210. doi:10.1021/ef200765f

Hicks, J. C., Drese, J. H., Fauth, D. J., Gray, M. L., Qi, G., & Jones, C. W. (2008). Designing adsorbents for CO_2 capture from flue gas-hyperbranched aminosilicas capable of capturing CO_2 reversibly. *Journal of the American Chemical Society*, *130*(10), 2902–2903. doi:10.1021/ja077795v PMID:18281986

Hori, K., Higuchi, T., Aoki, Y., Miyamoto, M., Yasunori, O., Yogo, K., & Uemiya, S. (2017). Effect of pore size, aminosilane density and aminosilane molecular length on CO_2 adsorption performance in aminosilane modified mesoporous silica. *Microporous and Mesoporous Materials*, *246*, 158–165. doi:10.1016/j.micromeso.2017.03.020

Huirache-Acuña, R., Nava, R., Peza-Ledesma, C. L., Lara-Romero, J., Alonso-Núñez, G., Pawelec, B., & Rivera-Muñoz, E. M. (2013). SBA-15 Mesoporous silica as catalytic support for hydrodesulfurization catalysts–Review. *Materials (Basel)*, *6*(9), 4139–4167. doi:10.3390/ma6094139 PMID:28788323

Jing, Y., Wei, L., Wang, Y., & Yu, Y. (2014). Synthesis, characterization and CO_2 capture of mesoporous SBA-15 adsorbents functionalized with melamine-based and acrylate-based amine dendrimers. *Microporous and Mesoporous Materials*, *183*, 124–133. doi:10.1016/j.micromeso.2013.09.008

Jung, H., Lee, C. H., Jeon, S., Jo, D. H., Huh, J., & Kim, S. H. (2016). Effect of amine double-functionalization on CO_2 adsorption behaviors of silica gel-supported adsorbents. *Adsorption*, *22*(8), 1137–1146. doi:10.1007/s10450-016-9837-2

Khatri, R. A., Chuang, S. S. C., Soong, Y., & Gray, M. (2005). Carbon dioxide capture by diamine-grafted sba-15: A combined fourier transform infrared and mass spectrometry study. *Industrial & Engineering Chemistry Research*, *44*(10), 3702–3708. doi:10.1021/ie048997s

Khatun, R., Bhanja, P., Molla, R. A., Ghosh, S., Bhaumik, A., & Islam, S. M. (2017). Functionalized SBA-15 material with grafted CO_2H group as an efficient heterogeneous acid catalyst for the fixation of CO_2 on epoxides under atmospheric pressure. *Molecular Catalysis*, *434*, 25–31. doi:10.1016/j.mcat.2017.01.013

Lakhi, K. S., Cha, W. S., Choy, J. H., Al-Ejji, M., Abdullah, A. M., Al-Enizi, A. M., & Vinu, A. (2016). Synthesis of mesoporous carbons with controlled morphology and pore diameters from SBA-15 prepared through the microwaveassisted process and their CO_2 adsorption capacity. *Microporous and Mesoporous Materials*, *233*, 44–52. doi:10.1016/j.micromeso.2016.06.040

Leung, Y. C. D., Caramanna, G., & Maroto-Valer, M. M. (2014). An overview of current status of carbondioxide capture and storage technologies. *Renewable & Sustainable Energy Reviews*, *39*, 426–443. doi:10.1016/j.rser.2014.07.093

Linfang, W., Lei, M., Aiqin, W., Gian, L., & Tao, Z. (2007). CO_2 Adsorption on SBA-15 Modified by Aminosilane. *Chinese Journal of Catalysis, 28*(9), 805–810. doi:10.1016/S1872-2067(07)60066-7

Linfang, W., Lei, M., Aiqin, W. Q. L., & Tao, Z. (2007). CO_2 adsorption on SBA-15 modified by aminosilane. *Chinese Journal of Catalysis, 28*(9), 805–810. doi:10.1016/S1872-2067(07)60066-7

Liu, Y., Lin, X., Wu, X., Liu, M., Shi, R., & Yu, X. (2017). Pentaethylenehexamine loaded SBA-16 for CO_2 capture from simulated flue gas. *Powder Technology, 318*, 186–192. doi:10.1016/j.powtec.2017.06.002

Loganathan, S., Tikmani, M., Edubilli, S., Mishra, A., & Ghoshal, A. K. (2014). CO_2 adsorption kinetics on mesoporous silica under wide range of pressure and temperature. *Chemical Engineering Journal, 256*, 1–8. doi:10.1016/j.cej.2014.06.091

Myers, D. (1992). Surfactant Science and Technology (2nd ed.). Academic Press.

Øye, G., Sjöblom, J., & Stöcker, M. (2001). Synthesis, characterization and potential applications of new materials in the mesoporous range. *Advances in Colloid and Interface Science, 89-90*, 439–466. doi:10.1016/S0001-8686(00)00066-X PMID:11215809

Samiey, B., Cheng, C. H., & Wu, J. (2014). Organic-inorganic hybrid polymers as adsorbents for removal of heavy metal ions from solutions: A Review. *Materials (Basel), 7*(2), 673–726. doi:10.3390/ma7020673 PMID:28788483

Sanz, R., Calleja, G., Arencibia, A., & Sanz-Pérez, E. S. (2012). Amino functionalized mesostructured SBA-15 silica for CO_2 capture: Exploring the relation between the adsorption capacity and the distribution of amino groups by TEM. *Microporous and Mesoporous Materials, 158*, 309–317. doi:10.1016/j.micromeso.2012.03.053

Sanz-Pérez, E. S., Murdock, C. R., Didas, S. A., & Jones, C. W. (2016). Direct capture of CO_2 from ambient air. *Chemical Reviews, 116*(19), 11840–11876. doi:10.1021/acs.chemrev.6b00173 PMID:27560307

Sanz-Pérez, E. S., Olivares-Marín, M., Arencibia, A., Sanz, R., Calleja, G., & Maroto-Valer, M. M. (2013). CO_2 adsorption performance of amino-functionalized SBA-15 under post-combustion conditions. *International Journal of Greenhouse Gas Control, 17*, 366–375. doi:10.1016/j.ijggc.2013.05.011

Shah, P., Ramaswamy, A. V., Lazar, K., & Ramaswamy, V. (2007). Direct hydrothermal synthesis of mesoporous Sn-SBA-15 materials under weak acidic conditions. *Microporous and Mesoporous Materials, 100*(1-3), 210–226. doi:10.1016/j.micromeso.2006.10.042

Shakerian, F., Kim, K. H., Szulejko, J. E., & Park, J. W. (2015). A comparative review between amines and ammonia as sorptive media for post-combustion CO_2 capture. *Applied Energy, 148*, 10–22. doi:10.1016/j.apenergy.2015.03.026

Thunyaratchatanon, C., Luengnaruemitchai, A., Chaisuwan, T., Chollacoop, N., Chen, S. Y., & Yoshimura, Y. (2017). Synthesis and characterization of Zr incorporation into highly ordered mesostructured SBA-15 material and its performance for CO_2 adsorption. *Microporous and Mesoporous Materials, 253*, 18–28. doi:10.1016/j.micromeso.2017.06.015

Ullah, R., Atilhan, M., Aparicio, S., Canlier, A., & Yavuz, C. T. (2015). Insights of CO_2 adsorption performance of amine impregnated mesoporous silica (SBA-15) at wide range pressure and temperature conditions. *International Journal of Greenhouse Gas Control, 43*, 22–32. doi:10.1016/j.ijggc.2015.09.013

Unveren, E. E., Monkul, B. O., Sarıoğlan, Ş., Karademir, N., & Alper, E. (2017). Solid amine sorbents for CO_2 capture by chemical adsorption: A review. *Petroleum, 3*(1), 37–50. doi:10.1016/j.petlm.2016.11.001

Wang, X., & Song, C. (2012). Temperature-programmed desorption of CO_2 from polyethylenimine-loaded SBA-15 as molecular basket sorbents. *Catalysis Today, 194*(1), 44–52. doi:10.1016/j.cattod.2012.08.008

Wang, X., & Song, C. (2014). New strategy to enhance CO_2 capture over a nanoporous polyethylenimine sorbent. *Energy & Fuels, 28*(12), 7742–7745. doi:10.1021/ef501997q

Wei, J., Shi, J., Pan, H., Zhao, W., Ye, Q., & Shi, Y. (2008). Adsorption of carbon dioxide on organically functionalized SBA-16. *Microporous and Mesoporous Materials, 116*(1-3), 394–399. doi:10.1016/j.micromeso.2008.04.028

Yan, X., Komarneni, S., & Yan, Z. (2013). CO_2 adsorption on Santa Barbara Amorphous-15 (SBA-15) and amine-modified Santa Barbara Amorphous-15 (SBA-15) with and without controlled microporosity. *Journal of Colloid and Interface Science, 390*(1), 217–224. doi:10.1016/j.jcis.2012.09.038 PMID:23084869

Yaumi, A. L., Abu Bakar, M. Z., & Hameed, B. H. (2017). Recent advances in functionalized composite solid materials for carbon dioxide capture. *Energy*, *124*, 461–480. doi:10.1016/j.energy.2017.02.053

Zhang, S. X., Mab, T. Y., Renc, T. Z., & Yuan, Z. Y. (2011). Zirconium ribonucleotide surface-functionalized mesoporous SBA-15 materials with high capacity of CO2 capture. *Chemical Engineering Journal*, *171*(1), 368–372. doi:10.1016/j.cej.2011.04.021

Zhang, X., Qin, H., Zheng, X., & Wu, W. (2013). Development of efficient amine-modified mesoporous silica SBA-15 for CO_2 capture. *Materials Research Bulletin*, *48*(10), 3981–3986. doi:10.1016/j.materresbull.2013.06.011

Zhao, D., Feng, J., Huo, Q., Melosh, N., Fredrickson, G. H., Chmelka, B. F., & Stucky, G. D. (1998). Triblock copolymer syntheses of mesoporous silica with periodic 50 to 300 angstrom pores. *Science*, *279*(5350), 548–552. doi:10.1126/science.279.5350.548 PMID:9438845

Zhihui, H., Donghui, Z., & Jixiao, W. (2011). Direct synthesis of amine-functionalized mesoporous silica for CO_2 adsorption. *Chinese Journal of Chemical Engineering*, *19*(3), 386–390. doi:10.1016/S1004-9541(09)60225-1

Zhou, J., Zhao, H., Li, J., Zhu, Y., Hu, J., Liu, H., & Hu, Y. (2013). CO_2 capture on micro/mesoporous composites of (zeolite A)/(MCM-41) with Ca^{+2} located: Computer simulation and experimental studies. *Solid State Sciences*, *24*, 107–114. doi:10.1016/j.solidstatesciences.2013.07.008

Chapter 8
Water and Water Security

Nadiye Gür
Mersin University, Turkey

ABSTRACT

Today, there are many studies about the problems that may be faced in the context of World Water Day. In this chapter, the structure, pollution, quality grading, and human health effects of water; possible pollution prevention measures; and water safety are discussed. It is expected that the world population, which is about 7 billion currently, will rise to 9 billion by 2050. Water consumption is expected to increase at a higher rate, which is a major problem for the environment. By 2025, it has been estimated that two-thirds of the world's population will deal with water shortage. The world is not as rich in water as once thought and, hence, is at high risk for water shortage. For these reasons, we must all fulfill our responsibility to leave a habitable world to future generations.

INTRODUCTION

A majority of the world when viewed from outer space is blue which represents to water (Cosgrove et al., 2014), (Kim et al., 2017). Approximately 70% of the Earth's surface is covered with water (Cowan et al., 2014). However, 97% of this water is seawater, which is very difficult to use directly because of its salt content. (Das et al., 2014). Of the remaining 3%, a majority of two-third is sequestered in ice (Obbard et al., 2014). For this reason, only 1% of the world's water is available for human consumption (Hoekstra, 2014).

Water is a fundamental substance of great importance to all plants, animals (Weil et al., 2016). All living things being in the world need water to grow and maintain

DOI: 10.4018/978-1-5225-3379-5.ch008

their existence (Black, 2016), as water comprises a large part of living things (De Asúa et al., 2017).

Water molecules are polar, leading to high solubility of salts in water (Snyder et al., 2014). In this regard, water creates a solution environment in which nutrients can be transported and biological reactions can occur in living cells (Tan, 2014). Many functions occur in our bodies. Owing to be a very good solvent water carries the nutrients necessary for our bodies to our cells and removes harmful substances from our cells (Hall, 2015). On the other hand, the fact that water is a very good solventthatcan cause to many deadly substances (Gschwend, 2016), harmful microorganisms, and water containing organic and inorganic chemicals unsuitable for human consumption (Priac et al., 2017) enter into human body. Water can be contaminated via (Spellman, 2017):

- The detergents we use and the kitchen dishwashing water, composed of dissolved organics, we discharge are polluting water resources. Many foods we use in our daily lives can have a dangerous effect when mixed with water. For example, if the sugar that we put in our tea mixes with a natural water source, it becomes a food source for the bacteria therein. Bacteria consume oxygen in the water along with the supplied nutrients, thus other aquatic organisms are prevented from consuming that oxygen and death may occur.
- Human and animal wastes, pesticides, and industrial plants that do not have wastewater treatment systems also damage water resources.

According to the water pollution control regulations set by the World Health Organization, waters are divided into four sections: high quality, less polluted, polluted, and heavily polluted (Zhang, 2017). Drinking water should be of high quality (Gazan et al., 2016). The water quality classes are based on physical, inorganic, organic, and bacteriological parameters (Gomes et al., 2014).

Some of the short-term effects to human health from dangerous substances in water include vomiting, fever, and dysentery. In the long run, death is a possibility (Raja, 2015). The highest number of deaths worldwide is due to consumption of unhealthy water (Blaikie et al., 2014).

World Water Day: Water and Food Security

More than 1 billion people in the world suffer water problems (Mekonnen et al., 2016). Women and children are most affected by this situation (Lawn et al., 2016). The water crisis affects many areas, such as education, health, and the work force, in a negative way (Brownson et al., 2017; Livi-Bacci, 2001).

In 1993, The United Nations (UN) General Assembly has declared March 22 as World Water Day to increase public awareness by the UN and its Member States regarding the protection of water resources (http://www.worldwaterday.org/) (Gleick, 1996). Every year a different theme is addressed at World Water Day events. States organize seminars and expositions in national context to inform public about the importance of freshwater and sustainable usage of fresh water resources (Chapman, 1992) (http://www.worldwaterday.org/). Themes of World Water Day from 1994 to 2020 are:

- 2020: Climate Change
- 2019: Leaving No One Behind (Human Rights and Refugees)
- 2018: Nature for Water
- 2017: Wastewater
- 2016: Water and Jobs
- 2015: Water and Sustainable Development
- 2014: Water and Energy
- 2013: Water Cooperation
- 2012: Water and Food Security
- 2011: Water for Cities
- 2010: Water Quality
- 2009: Transboundary Waters
- 2008: International Year of Sanitation
- 2007: Water Scarcity – Coordinator: FAO
- 2006: Water and Culture
- 2005: Water for Life 2005-2015
- 2004: Water and Disasters
- 2003: Water for the Future
- 2002: Water for Development
- 2001: Water for Health, Taking Charge
- 2000: Water for the 21st Century
- 1999: Everyone lives downstream
- 1998: Groundwater, the invisible resource
- 1997: The World's Water, Is There Enough?
- 1996: Water for Thirsty Cities
- 1995: Women and Water
- 1994: Caring for Water Resources is Everybody's Business (http://www.unwater.org/what-we-do/inspire-action/)

The theme of 2012 was "Water and Food Security" (Keesstra et al., 2016). Water is used very differently for agricultural and home applications (Black, 2016), and 70% of the world's water is consumed via agricultural activities (Alcamo et al., 20000).

Currently 7 billion people needs food in the World. Population is expected to be 9 billion in 2050. Each human drinks water 2-4 liters per day; on the other hand most of the water consumed is hidden in the food. (Pan, 2013). 2,000-5,000 liters of water is "eaten" daily based on the production processes of the food (Nedellec et al., 2016). For example, the volume of water needed to produce some everydayconsumables include:

- 70 liters for 1 apple,
- 135 liters for 1 egg,
- 140 liters for 1 cup of coffee,
- 170 L for 1 glass of orange juice,
- 200 L for 1 glass of milk, and
- 7000 L water is consumed for 1 steak.

Currently, 1 billion people are suffering chronic starvation (Banik, 2016). Water resources are under great pressure (Leng, 2015). We all have a duty to ensure the increasing population's food and water needs are met. Some measures that can be taken to achieve this are:

- Following healthier and sustainable diet.
- Consuming less-water intensive products.
- Reducing the food wastage (30% of the food produced worldwide is not used; water used for these food are lost).
- Producing higher quality food with less water (http://www.unwater.org/what-we-do/inspire-action/).
- Reducing unnecessary water consumption at home.
- Changing architectural planning in such a way that; 1-two cuvette must be made one is used for washing vegetables, fruits, etc. to produce clean waste water that can be used for various purpose, such as irrigation of plants in the garden, washing cars etc. 2-the other one must be used to wash very dirty materials, waste water from dishwasher discharged to sewerage system.
- Collecting rain water, the distilled water quality except some particles coming down from the atmosphere and discharging to a container that can be used for washing machine that reduces the amount of detergents, some salts used for softening of water.

- Discharging bath water, containing mainly soap, separately and collecting in a container (by the addition of lime to settle the hard soap and the water left can be used for plants in the garden or around).

CONCLUSION

It is expected that the world population, which is about 7 billion currently, will rise to 9 billion by 2050. Water consumption is expected to increase at a higher rate, which is a major problem for the environment. By 2025, it has been estimated that two-thirds of the world's population will deal with water shortage. The world is not as rich in water as once thought and, hence, is at high risk for water shortage. For these reasons, we must all fulfill our responsibility to leave a habitable world to future generations.

Some suggestions for keeping water resources clean are as follows.

- It is very important that we keep the sources of water clean. To this end, you should consume as little water as possible and introduce as few harmful components as possible to the water. For example, dishes should be washed after physically separating food residues from them. Thus, fewer organic components are sent to the sewer system.
- Farming and settlement in water basins should not be permitted.
- People should be made aware that detergents should be used to clean oils only and no positive results occur from using greater amounts of detergent in the kitchen and laundry
- Agricultural activities should be carried out in environmentally friendly ways

In order to meet the food and water need of growing population every person should contribute. Appropriate measures are:

- Adopting a healthier, sustainable diet;
- Consuming foods that use low volumes of water in production;
- Preventing food waste (30% of food worldwide is wasted and, thus, the water used to produce it is lost);
- Producing higher quality food with less water;
- Changing architectural planning to sustain re-use of water;
- Collecting and using rain water.

REFERENCES

Alcamo, J., Henrichs, T., & Rösch, T. (2000). *World Water in 2025—Global modeling and scenario analysis for the World Commission on Water for the 21st Century, Kassel World Water Series 2*. Center for Environmental Systems Research, University of Kassel. http://www.usf.uni-kassel.de/usf/archiv/dokumente.en.htm

Banik, D. (2016). The Hungry Nation: Food Policy and Food Politics in India. *Food Ethics, 1*(1), 29–45. doi:10.1007/s41055-016-0001-1

Black, M. (2016). *The Atlas of Water: Mapping the World's Most Critical Resource*. Univ. of California Press.

Blaikie, P., Cannon, T., Davis, I., & Wisner, B. (2014). *At Risk: Natural Hazards, People's Vulnerability and Disasters*. Routledge.

Brownson, R. C., Baker, E. A., Deshpande, A. D., & Gillespie, K. N. (2017). *Evidence-Based Public Health*. Oxford University Press.

Chapman, D. (Ed.). (1992). *Water quality assessments. A guide to the use of biota, sediments and water in environmental monitoring*. London: Chapman & Hall. doi:10.4324/9780203476710

Cosgrove, W. J., & Rijsberman, F. R. (2014). *World Water Vision: Making Water Everybody's Business*. Routledge.

Cowan, N. B., & Abbot, D. S. (2014). Water cycling between ocean and mantle: Super-Earths need not be waterworlds. *The Astrophysical Journal, 781*(1), 27. doi:10.1088/0004-637X/781/1/27

Das, R., Ali, E., Hamid, S. B. A., Ramakrishna, S., & Chowdhury, Z. Z. (2014). Carbon nanotube membranes for water purification: A bright future in water desalination. *Desalination, 336*, 97–109. doi:10.1016/j.desal.2013.12.026

De Asúa, M., & French, R. (2017). *A New World of Animals: Early Modern Europeans on the Creatures of Iberian America*. Routledge.

Energy Saving Trust. (2013). Retrieved from http://www.energysavingtrust.org.uk/sites/default/files/reports/AtHomewithWater%287%29.pdf

Gazan, R., Sondey, J., Maillot, M., Guelinckx, I., & Lluch, A. (2016). Drinking Water Intake Is Associated with Higher Diet Quality among French Adults. *Nutrients, 8*(11), 689. doi:10.3390/nu8110689 PMID:27809236

Gleick, P. (1996). Basic water requirements for human activities: Meeting basic needs. *Water International, 21*(2), 83–92. doi:10.1080/02508069608686494

Gomes, A., Pires, J., Figueiredo, S., & Boaventura, R. (2014). Optimization of river water quality surveys by multivariate analysis of physicochemical, bacteriological and ecotoxicological data. *Water Resources Management*, *28*(5), 1345–1361. doi:10.1007/s11269-014-0547-9

Gschwend, P. M. (2016). *Environmental Organic Chemistry*. John Wiley & Sons.

Hall, J. E. (2015). *Guyton and Hall Textbook of Medical Physiology E-Book*. Elsevier Health Sciences.

Hoekstra, A. Y. (2014). Sustainable, efficient, and equitable water use: The three pillars under wise freshwater allocation. *Wiley Interdisciplinary Reviews: Water*, *1*(1), 31–40. doi:10.1002/wat2.1000

Keesstra, S., Quinton, J., van der Putten, W., Bardgett, R., & Fresco, L. (2016). The significance of soils and soil science towards realization of the United Nations Sustainable Development Goals. *Soil (Göttingen)*, *2*(2), 111–128. doi:10.5194/soil-2-111-2016

Kim, J., Son, S.-W., Gerber, E., & Park, H.-S. (2017). Defining Sudden Stratospheric Warming in Climate Models: Accounting for Biases in Model Climatologies. *Journal of Climate*, *30*(14), 5529–5546. doi:10.1175/JCLI-D-16-0465.1

Lawn, J. E., Blencowe, H., Waiswa, P., Amouzou, A., Mathers, C., Hogan, D., & Shiekh, S. (2016). Stillbirths: Rates, risk factors, and acceleration towards 2030. *Lancet*, *387*(10018), 587–603. doi:10.1016/S0140-6736(15)00837-5 PMID:26794078

Leng, G., Huang, M., Tang, Q., & Leung, L. R. (2015). A modeling study of irrigation effects on global surface water and groundwater resources under a changing climate. *Journal of Advances in Modeling Earth Systems*, *7*(3), 1285–1304. doi:10.1002/2015MS000437

Livi-Bacci, M. (2001). *A Concise History of World Population* (3rd ed.). Blackwell.

Mekonnen, M. M., & Hoekstra, A. Y. (2016). Four billion people facing severe water scarcity. *Science Advances*, *2*(2), 1–6. doi:10.1126/sciadv.1500323 PMID:26933676

Nedellec, V., Rabl, A., & Dab, W. (2016). Public health and chronic low chlordecone exposures in Guadeloupe; Part 2: Health impacts, and benefits of prevention. *Environmental Health*, *15*(1), 78. doi:10.1186/s12940-016-0159-3 PMID:27430869

Obbard, R., Sadri, S., Wong, Y. Q., Khitun, A. A., Baker, I., & Thompson, R. C. (2014). Global warming releases microplastic legacy frozen in Arctic Sea ice. *Earth's Future*, *2*(6), 315–320. doi:10.1002/2014EF000240

Pan, A., Malik, V. S., Hao, T., Willett, W. C., Mozaffarian, D., & Hu, F. B. (2013). Changes in water and beverage intake and long-term weight changes: Results from three prospective cohort studies. *International Journal of Obesity*, *37*(10), 1378–1385. doi:10.1038/ijo.2012.225 PMID:23318721

Priac, A., Morin-Crini, N., Druart, C., Gavoille, S., Bradu, C., Lagarrigue, C., & Crini, G. et al. (2017). Alkylphenol and alkylphenolpolyethoxylates in water and wastewater: A review of options for their elimination. *Arabian Journal of Chemistry*, *10*, S3749–S3773. doi:10.1016/j.arabjc.2014.05.011

Raja, A. A. (2015). *Study of acute toxicity of an unani medicine used in dysentery. Unpublished doctoral dissertation*. BRAC University. Retrieved from http://www.wateraidamerica.org/sites/default/files/attachments/Rainwater%20harvesting.pdf

Snyder, P., Lockett, M., Moustakas, D., & Whitesides, G. (2014). Is it the shape of the cavity, or the shape of the water in the cavity? *The European Physical Journal. Special Topics*, *223*(5), 853–891. doi:10.1140/epjst/e2013-01818-y

Spellman, F. R. (2017). *The Drinking Water Handbook*. CRC Press. doi:10.1201/9781315159126

Tan, K. (2014). *Humic Matter in Soil and the Environment: Principles and Controversies*. CRC Press, Taylor & Francis Group.

Weil, R. R., & Brady, N. C. (2016). *The Nature and Properties of Soils*. Columbus, OH: Pearson.

Zhang, Y., Chu, C., Li, T., Xu, S., Liu, L., & Ju, M. (2017). A water quality management strategy for regionally protected water through health risk assessment and spatial distribution of heavy metal pollution in 3 marine reserves. *The Science of the Total Environment*, *599*, 721–731. doi:10.1016/j.scitotenv.2017.04.232 PMID:28499221

Chapter 9
Water Pollution Burden and Techniques for Control

Kanav Dhir
DAV College, India

Meenakshi Jatayan
PEC University of Technology, India

Shakti Kumar
PEC University of Technology, India

ABSTRACT

The enhancements in the socio-economic status of many people has come from the expansion of agricultural and industrial production. But, some of the activities associated with this expansion have adversely affected water quality. This leads to a negative impact on public health, eminence of life, and environment. This chapter sets out to explain the various factors that lead to water contamination and different mitigation techniques to manage them. We need this knowledge so as to develop suitable solutions for a broad range of environmental problems.

INTRODUCTION

Water is the basic need for sustaining life as its availability both in quantity and quality are important. Earlier importance of the water was from the viewpoint of the quantity as civilization developed around the water bodies for the support of agriculture, transportation and domestic purposes. In the past, quality of water was judged through physical senses of sight, taste and smell. Recent development in medical sciences suggests some other methods for determination of water quality on

DOI: 10.4018/978-1-5225-3379-5.ch009

the basis of biological, chemical and pathological pathways (Aggarwal, 2005). The relationship between human waste, drinking water and diseases (Halder et al, 2015) that are transmitted to humans through drinking contaminated water is found and is shown in Table 1. With the rapid increase in industrialisation and population, the demand for water is shooting up every day. Clean drinking water supply and basic sanitation are vital human needs for good health. Water performs several metabolic, physiological and other essential activities in human body and other living beings. It is used in our daily activities for various purposes such as drinking, cooking, bathing and in washing clothes and utensils.

Water is the most abundant compound found in the nature covering 70.9% of the earth surface and is vital for all known forms of life. Oceans hold 97% of surface water, polar ice caps contain 2.4% and land surface water such as rivers, lakes and ponds have 0.6% of the total water. Ground water makes up about 20% of the world's fresh water supply, which is about 0.61% of the entire world's water including ocean and ice. Global ground water storage is roughly equal to the total amount of fresh water stored in the snow and ice pack including the north and south poles. According to the UN, planet earth mean annual renewable volume of water is 43,000 cubic kilometres. This is about half of all the fresh water contained in all the earth natural lakes and about ten times the volume of all man-made reservoirs. Groundwater recharge accounts for about 10,000 cubic kilometres annually (i.e. 0.1% of all ground water resources). Thus, only a small proportion of the total volume of groundwater reserves is recharged every year as compared to the total large volume in stock. Some groundwater systems are non-renewable under current climatic conditions because they are formed under much wetter climates that prevailed perhaps 1000 or 10,000 years ago. Groundwater acts as a conduit which can transport water over long distances. It acts as a mechanical filter which improves water quality by removing suspended solids and bacterial contamination. The sources of groundwater (fresh water) are wells and springs which are also the basic source for rural domestic use. Groundwater reservoirs are being increasingly mined in the arid zones of the world. It is replenished by precipitation through rain, snow, sleet and hail.

Water present on earth is in the constant state of motion through a cycle of evaporation and transpiration (evapo-transpiration). The fresh water on the earth exists in surface water bodies such as river and lakes, glaciers and ground water. But these sources of fresh water are shrinking and majority of them are getting polluted by anthropogenic factors such as discharge of domestic wastes and industrial effluents (Mukherjee et al, nd). Excessive extraction of groundwater is also making the water a scarce commodity. The suitability of water in terms of public health is determined by physical, chemical, microbiological and radiological characteristics. Out of these, the most important are chemical and microbiological quality. There are many ways in which water pollution can be defined and usually it means the contamination of

Table 1. Common diseases transmitted to humans through drinking contaminated water

Types of organisms	Diseases	Effects
Bacteria	Typhoid fever	• Diarrhoea • Severe vomiting • Enlarged spleen • Inflamed intestine • Often fatal if untreated
	Cholera	• Diarrhoea • Severe vomiting • Dehydration • Often fatal if untreated
	Bacterial dysentery	• Diarrhoea • Bleeding • rarely fatal except in infants without proper treatment
	Enteritis	• Severe stomach pain • Nausea • Vomiting • Rarely fatal
Viruses	Infectious hepatitis (Type B)	• Fever • Severe headache • Loss of appetite • Abdominal pain • Jaundice • Enlarged liver • Rarely fatal but may cause permanent liver damage
	Poliomyelitis	• Fever • Diarrhoea • Backache • Sore throat • Aches in limbs • Can infect spinal cord and cause paralysis • Muscle weakness
Parasitic protozoa	Amoebic dysentery	• Severe diarrhoea • Headache • Abdominal pain • Chills • Fever; if not treated can cause liver abscess, bowel perforation and death
	Giardiasis	• Diarrhoea • Abdominal cramps • Flatulence • Belching • Fatigue
	Crytosporidium	• Severe diarrhoea • Cramps for up to 3 weeks • Possible death for people with weakened immune systems
Parasitic worms	Schistosomiasis	• Abdominal pain • Skin rash • Anaemia • Chronic fatigue • Chronic general ill health
	Ancylostomiasis	• Severe anaemia • Possible symptoms of bronchial infection

water bodies like groundwater, aquifers, lakes, rivers and oceans. Contamination occurs when pollutants are directly or indirectly discharged into water bodies. For example, half of the World Rivers are polluted due to contamination of toxic inorganic and organic chemicals caused by industries and mines as shown in Figure 1.

Water bodies can naturally clean certain amount of pollution by dispersion but it fails when large amount of untreated waste is dumped. This in turn affects the health of all the organisms which are dependent on that water body. Main source of water pollution is anthropogenic, which changes the quality and ecological status of water. Surface water pollution is categorized into point and non-point sources on the basis of their origin. But, ground water cannot be classified in to these above said categories. Chemicals may enter waterways from a point or a non-point source. Point source pollution (Figure 2(a)) is caused from discharges by a single source, specific site or an industrial site which is easy to identify, monitor and regulate. Non-point source pollution (Figure 2(b)) involves many small sources that combine to cause significant pollution which is difficult to identify and is expensive to clean

Figure 1. Images of polluted river water: (a) raw sewage in Iraq; (b) polluted river in China; (c) truck disposing of garbage into a Peru River; and (d) Ganga river in India

Figure 2. Water pollution: (a) Point Source of polluted water in gargas; (b) Non-point deposit from unprotected farmland flows into streams

(a) (b)

up. For e.g., the movement of rain or irrigation water picks up pollutants such as fertilizers and insecticides and bring them into rivers, lakes, reservoirs, coastal waters or groundwater. Another non-point source is storm water that collects on roads and eventually reaches rivers or lakes. Release of chemicals, contaminants into the soil located far away from the surface water body does not create point or non-point pollution but it can contaminate the aquifer present below. This create toxic plume which travels along the hydrological gradient and contaminate other aquifers along its pathway. Chemical pollutants, pathogens, physical changes (elevated temperature and discoloration), change in substance concentration (i.e. calcium, sodium, etc.) and organic content leads to depletion of oxygen which has negative impact on aquatic flora and fauna.

2. USES OF WATER

Uses of water in different fields (Aggarwal, 2015) are enlisted as follows:

1. The most important use of water is in agriculture for irrigation (Mukherjee et al, nd).
2. Human body contains water anywhere from 55% to 78% depending on the body size. The body requires it in between one to seven liters of water per day to stay healthy. The precise amount of water required depends on the level of activity, temperature, humidity and so on.

3. The propensity of water to form solutions and emulsions is useful in various washing processes. Washing is also an important component of several aspects of personal body hygiene.
4. Water is widely used in chemical reactions as a solvent or reactant.
5. Water and steam are used as heat transfer fluids in diverse heat exchange systems due to its easy availability and high heat capacity. Many electric power stations use water as coolant which vaporizes and drives steam turbines to drive generators. In nuclear power industry, water is used as a neutron moderator.
6. Water has high heat of vaporization which makes it a good fire extinguishing fluid as evaporation of water carries heat away from the fire.
7. Humans use water for many recreational purposes such as swimming, waterskiing, boating, surfing and diving. In addition, some sports like ice hockey and ice skating are played on ice.
8. The water industry provides drinking water and wastewater services (including sewage treatment) to households and industry. Water supply facilities include water wells cisterns for rainwater harvesting, water supply network, water purification facilities, water tanks, water towers, water pipes including old aqueducts (Figure 3).
9. Water is used in power generation. Hydroelectricity is electricity obtained from hydropower. Hydroelectricity is a low-cost, non-polluting and renewable energy source.

Figure 3. Uses of water in our daily life

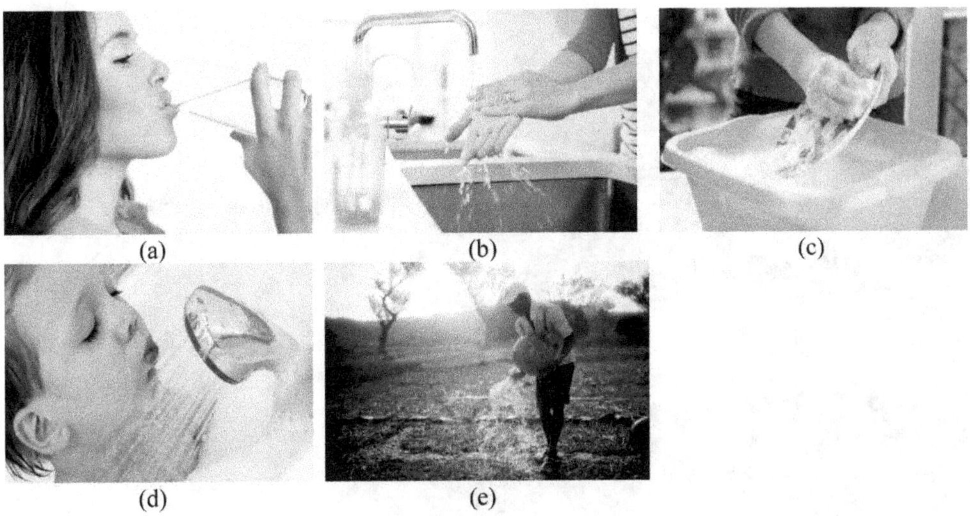

(a) (b) (c) (d) (e)

3. WATER POLLUTION

Nature is beyond the control of human being but it is being affected by their unnecessary activities that are running behind the illusion of rapid growth of advancement (Halder et al, 2015). There are two main categories that explain the reasons for water pollution.

1. **Natural Sources:** Sometime nature itself become the cause of pollution for example: eruption of volcano degraded the water resources with its fly ashes and magma, arrival of floods that decline the quality of water and many more.
2. **Anthropogenic Factors:** These factors are associated with various man-made activities that cause water pollution (Goel et al, nd; Peavy et al, 1985) and are enlisted as follows:
3. **Thermal Pollutants:** Rate of biological activity increases with the rise of the temperature which decreases the amount of dissolved oxygen. It leads to the scarcity of oxygen that affects the growth of aquatic organisms in water bodies (Figure 4). These detrimental effects are as follows:
    a. Reduction in dissolved oxygen
    b. Increase in BOD
    c. Excessive eutrophication
    d. Decrease in solubility of gases in water
    e. Rapid setting of sediment load in water affecting aquatic food supply

Figure 4. Thermal pollutants

4. **Radioactive Materials in Water:** Radioactive pollutants enter into the water streams from various sources such as nuclear power plants, nuclear reactors and fusion products, etc. to which living organisms are adapted with various ill effects. Polluted water containing radioisotopes produces a set of syndromes characterized by nausea, vomiting, diarrhoea, epilation along with general weakness which are known as radiation sickness. It destroys biological immune system i.e. body becomes less resistant towards a variety of diseases. It also causes somatic and genetic disorders, gene mutations and blood abnormalities in higher animals including man. Radioactive elements present in water accumulate in soil sediments, air and aquatic ecosystems. Trace amounts of radionuclides may lead to increase in rate of mutation of plants.
5. **Sewage and Domestic Wastes:** Sewage is an excellent medium for the growth of pathogenic bacteria, viruses and protozoa. Domestic sewage which is primarily composed of spent water containing soapy water, food material, makes water completely unfit for drinking and domestic use. Several pathogenic microorganisms introduced into the water cause deleterious effects and produces chronic diseases in man and animals. Dirty water can have any type of germs, which produces water-borne diseases like jaundice, cholera, typhoid, etc. and thus, may affect human health.
6. **Industrial Effluent:** Industrial wastewater is contaminated by a variety of organic and inorganic pollutants (Figure 5). All the rivers of India are heavily polluted by the discharge of industrial wastewater. Due to the presence of toxic

Figure 5. Industrial Effluents

substances in the effluent it produces deleterious effects on living organisms and brings about death of sub lethal pathology of kidney, liver, brains, and lungs. Disinfectants which are added in water to control algae growth and bacteria may persist in water bodies and may cause mortality of fish. Effluents containing acids and alkalies make the water corrosive (Abdulraheem, 1989). Municipal wastewater also carries organic wastes which are biodegradable and this water is directly discharged into rivers which affect animal life. So, industrial and municipal wastewater is treated in Effluent Treatment Plant (ETP) prior to their disposal in rivers or water bodies

7. **Agricultural Discharge:** Pesticides, fertilizers, farm wastes, manure slurry, drainage from silage, plant and animal debris contains mostly the inorganic materials that causes heavy pollution to water resources (Figure 6).
8. **Fertilizers:** Excessive application of nitrogen fertilizers to the soil often leads to accumulation of nitrates in the water which when drunk by cattle and man get reduced to toxic nitrates by the intestinal bacteria. Nitrates enter the blood stream and react with haemoglobin, which has a stronger affinity for nitrates than for oxygen to form methamoglobin. This is called methaemogiobinemia or Mtte-baby syndrome. This causes damage to the respiratory and vascular systems, because of which suffocation and death may occur (Masters et al, 2007). Agricultural fertilizers crowd out essential nutrients present in the topsoil layers. The microbe-enriched humus enhances plant growth. But fertilizer

Figure 6. Pesticides causing water pollution

enriched soil cannot support the microbial life for a long time. Fertilizers used to increase the growth of crops also increase the algal growth in surface water into which they are washed. This increased water fertility causing algal and water plant growth is called eutrophication. It is the process of nutrient enrichment of water and result in loss of species diversity.

9. **Detergents:** Detergents are used as cleansing agents and derived from surfactants (10-30%), builder (15%) and other ingredients. Household detergents (Figure 7) contain several pollutants, which severely affect the water bodies.
10. **Toxic Metals:** Amongst the industries with the highest emission of heavy metals are the mining industry, metallurgical industry, leather industry, distilleries, battery industry and thermal power plants. Toxic metals are added in aquatic system from industrial processes and domestic sewage discharge (Masters et al, 2007; Halder 2015). Traces of heavy metals such as Hg, Cd, Pb, As, Co, Mn, Fe and Cr have been identified as deleterious to aquatic ecosystem and human health. Waste containing high concentration of toxic metals either separately or in combination is extremely toxic to all living organisms.
11. **Effect of Bioaccumulation:** Bioaccumulation is building up of the toxic substances in food chain. Mercury poisoning (Minamata disease) is caused due to consumption of fish captured from Minamata Bay in Japan. For example:

Figure 7. Detergents causing water pollution

accumulations of the heavy metals like mercury in fish. Mercury enters the chain by absorbing on the algae in form of methyl mercury. Fish eat the algae and absorb methyl mercury, since absorbance of mercury in fish is faster than excretion; it gets accumulated in the fish body. Further up the food chain, predatory fish and birds then absorb the mercury from the fish they consume, which then accumulates in their bodies leading to a higher concentration of the mercury in their own bodies than in the species they have eaten. This is known as bio-magnification (Aggarwal 2005; Goel, 2006). This process can be dangerous to humans as we could consume fish which have bio-accumulated mercury and absorb it ourselves, causing health problems such as damage to the central nervous system.

Major water pollutants and their sources are shown in Table 2.

Table 2. Major water pollutants and their sources

Types	Effects	Examples	Major sources
Infectious agents (pathogens)	Cause diseases	• Bacteria • Viruses • Protozoa • Parasites	• Human wastes • Animal wastes
Oxygen-demanding wastes	Deplete dissolved oxygen needed by aquatic species	• Biodegradable animal species • Plant debris	• Sewage • Animal feedlots • Food processing facilities • Pulp mills
Plant nutrients	Cause excessive growth of algae and other species	• Nitrates (NO_3^-) • Phosphates (PO_4^{3-})	• Sewage • Animal wastes • Inorganic fertilizers
Organic chemicals	Add toxins to aquatic systems	• Oil • Gasoline • Plastics • Pesticides • Cleaning solvents	• Industry • Farms • Households
Inorganic chemicals	Add toxins to aquatic systems	• Acids • Bases • Salts • Metal compounds	• Industry • Households • Surface runoff
Sediments	Disrupt photosynthesis, food webs, other processes	• Soil • Silt	Soil erosion
Heavy metals	Cause cancer, disrupt immune and endocrine systems	• Lead • Mercury • Arsenic	• Unlimited landfills • Household chemicals • Mining refuse • Industrial discharges
Thermal	Make some species vulnerable to disease	Heat	Electric power and industrial plants

4. TRANSPORTATION AND CHEMICAL REACTIONS OF POLLUTANTS

The river or stream system both produces and consumes oxygen. It takes oxygen from the atmosphere and plants as a result of photosynthesis. Running water because of its turbulence dissolves more oxygen than still water. Respiration by aquatic animals, decomposition and various chemical reactions consume oxygen.

Groundwater pollution (Rao et al, 1998) is much more difficult to decline than surfacewater pollution. This is because groundwater can move at great distances through unseen aquifers. Non-porous aquifers such as clays partially purify water by simple filtration (adsorption and absorption). Groundwater that moves through cracks and caverns is not filtered and can be transported as easily as surface water. In fact, this can be aggravated by the human tendency to use natural sinkholes as dumps in areas. Surface water (Ellis, 1989) is highly susceptible to contamination as compared to the groundwater.

Most water pollutants are eventually carried by rivers into the oceans. In some areas of the world the influence can be traced hundred miles from the mouth by studies using hydrology transport models. Many chemicals undergo reactive decay or chemically change especially over long periods of time in groundwater reservoirs. Such chemicals are the chlorinated hydrocarbons such as trichloroethylene (industrial metal degreasing and electronics manufacturing) and tetrachloroethylene used in the dry cleaning industry. Both of these chemicals are carcinogenic in nature and undergo partial decomposition reactions leading to new hazardous chemicals including dichloroethylene and vinyl chloride.

Waste water from sewage treatment plants often contains organic materials that are decomposed by microorganisms, which use oxygen in the process. The amount of oxygen consumed by these organisms in breaking down the waste is known as the biochemical oxygen demand (BOD). The chemical oxygen demand (COD) is the indirect measure of the amount of organic compounds in water. COD is less specific than BOD, since it measures everything that can be chemically oxidized rather than just levels of biodegradable organic matter. It helps in the determination of the organic compounds in the surface water like lakes and rivers or wastewater. Other sources of oxygen consuming waste include storm water runoff from farmland or urban streets, feedlots, and failing septic systems.

The amount of oxygen required, oxidizes organic compounds to carbon dioxide, ammonia and water is given by the equation.

$$C_nH_aO_bN_c+\left(n+\frac{a}{4}-\frac{b}{2}-\frac{3c}{4}\right)O_2\rightarrow nCO_2+\left(\frac{a}{2}-\frac{3c}{2}\right)H_2O+cNH_3$$

Oxygen found in dissolved form in water bodies is known as dissolved oxygen (DO). If more oxygen is consumed than produced, dissolved oxygen levels decline and some sensitive animals may move away, weaken, or may even die. DO vary with water temperature and altitude. Cold water holds more oxygen than warm water and water holds less oxygen at higher altitudes. Thermal discharges raise the temperature of water and lower down its oxygen content. Aquatic animals are most vulnerable to lowered DO levels in the early morning on hot summer days. This is because at that time stream flows are low, water temperatures are high and aquatic plants have not been producing oxygen since sunset. In contrast to lakes, where DO levels are most likely to vary vertically in the water column, the DO in rivers and streams changes more horizontally along the course of the waterway. This is especially true in smaller and shallower streams. In larger and deeper rivers, some vertical stratification of dissolved oxygen might occur. DO is measured either in milligrams per litre (mg/L) or "percent saturation". A milligram per litre is the amount of oxygen in a litre of water. Percent saturation is the amount of oxygen in a litre of water relative to the total amount of oxygen that the water can hold at that temperature. Dissolved oxygen is measured primarily either by using some variation of the Winkler method or by using a meter and probe.

$$Organic\,matter + O_2 \xrightarrow{micro-organism} CO_2 + H_2O + new\,cell + stable\,products$$

$$Organic\,matter \xrightarrow{micro-organism} CO_2 + H_2O + new\,cell + unstable\,products$$

The depletion in oxygen is measured by different methods such as BOD, which is further sub-divided into carbonaceous BOD (CBOD) and nitrogenous BOD (NBOD). Pollutants in the ground water are affected by the physical processes of advection, dispersion, sorption and volatization.

5. SURFACE WATER POLLUTION

Surface water differs from groundwater by having a stronger interaction with the atmosphere, a weaker interaction with sediments and by being penetrated by sunlight (Ellis, 1989). As a result, the nature and rate of the various bio-geochemical processes that act in surface water differ from those in groundwater. Generally, the redox potential is higher in surface water than in groundwater, because of the presence of relatively high concentrations of dissolved oxygen. The oxygen originates from diffusion of atmospheric oxygen or photosynthesis by primary producers (e.g. algae and submerged aquatic macrophytes) during daylight. Photosynthesis simultaneously extracts dissolved CO_2 (carbonic acid) from the water, which raises the pH. Therefore, surface waters are mostly neutral to basic, except for water bodies that are susceptible to acidification in poorly buffered catchments.

Compared to groundwater, surface water flows fast. Whereas groundwater travels at the rate of centimetres per day or even less, surface water travels at a rate in the order of metres to tens of metres per minute. Because of the fast transport rate of surface water, the reaction kinetics usually becomes apparent in the spatial variation in surface water composition. Accordingly, water pollutants that are discharged once off or continuously into surface waters are rapidly carried downstream and can assert their influence across distances up to hundreds of kilometres, as unfortunately demonstrated by large accidental spills into rivers.

5.1. Mitigation Techniques for Surface Water Pollution

The treatment of water can be divided into treatment of raw water for drinking purposes and treatment of wastewater for reuse (Ellis, 1989). Water used for household and domestic application should be free from microorganisms but can contain appreciable amount of dissolved salts. Whereas water used in industries can have microorganism but the water should be soft so that scaling of boilers doesnot take place. The method and degree used for treatment of water is site specific, it depends on the basis of type and level of waste; the disposal and usage of wastewater is decided. Wastewater is the water which is no longer suitable for the use and treatment or process to convert this water into reusable is called waste water treatment (Asolekar, nd). If wastewater is from municipal sources it is called sewage and treatment is called sewage treatment. Environmental public works deals with the waste water treatment along with the solid waste drainage management, sewage management and water management. By products from the wastewater treatment plants are also treated in the treatment plant. *Industrial wastewater* is the wastewater generated from the industrial and commercial areas. This wastewater contains objectionable organic and inorganic compounds that may not be amenable to conventional treatment processes. *Sewage* is the liquid waste originating from the domestic uses of water. It includes sullage, discharge from toilets, urinals, wastewater generated from commercial establishments, institutions, industrial establishments and also the groundwater and storm water that may enter into the sewers. Its decomposition produces large quantities of malodorous gases and it contains numerous pathogenic or disease producing bacteria, along with high concentration of organic matter and suspended solids. *Sewage Treatment Plant* is a facility designed to receive the waste from domestic, commercial and industrial sources and to remove materials that damage water quality and compromise public health and safety when discharged into water receiving systems or land. It is combination of unit operations and unit processes developed to treat the sewage to desirable standards to suit effluent norms defined by regulating authority. *Wastewater* includes both organic and inorganic constituents in soluble or suspended form and mineral content of liquid waste carried through liquid media. Generally the organic portion of the wastewater undergoes biological decompositions and the mineral matter may combine with water to form dissolved solids. *Sewerage* is the infrastructure which includes device, equipment and appurtenances for the collection, transportation and pumping of sewage, but excluding works for the treatment of sewage. Basically it is a water carriage system designed and constructed for collecting and carrying of sewage through sewers. Storm water indicates the rain water of the locality. *Subsoil water* is the groundwater that enters into the sewers through leakages is called subsoil water. *Sullage* is the wastewater generated from bathrooms, kitchens, washing place and wash basins, etc. Composition of this waste does not involve higher concentration of organic matter and it is less polluted water as compared to sewage.

Following are the mitigation techniques for surface water pollution:

1. **Raw Water Treatment:** For the treatment of the water used for drinking purpose, depends on the source of water. In some cases, single disinfection is all needed to use it for drinking whereas in other cases number of steps needed for the disinfection. In general practice, surface water is filtered and disinfected, while groundwater having hardness is removed by disinfection (Ellis, 1989).

The steps for raw water treatment (Figure 8) are explained as follows:

1. **Source Screening:** It is used to remove relatively large floating and suspended debris.
2. **Mixing Tank:** Mixing the matter with chemicals will encourage suspended solids to coagulate into larger particles and will get settled easily.

Figure 8. Typical treatment plant for surface water (addition step of softening is required in groundwater)

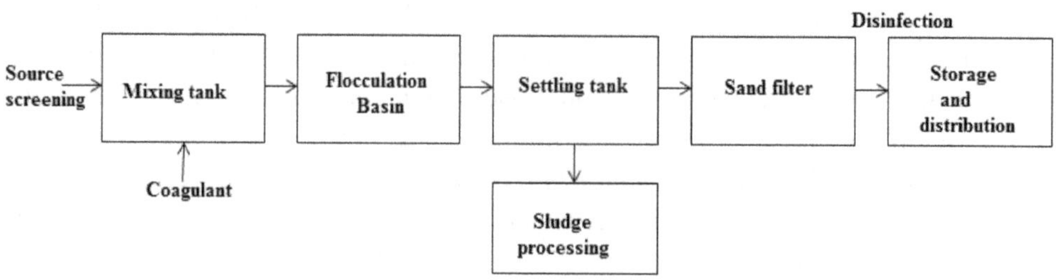

3. **Flocculation:** It is the process of gently mixing the water and coagulant, allowing the formation of large floc.
4. **Settling Tank:** In sedimentation, flow is reduced and gravity causes the floc to settle.
5. **Sludge Processing:** In sludge processing, mixture of solids and liquids get collected from the settling tank are dewatered and disposed off.
6. **Disinfection:** Disinfection of the liquid effluent to ensure that the water is free from harmful pathogens.
    a. **Industrial Water Treatment:** For designing and operating industrial water treatment facility, some factors are considered: water quantity is required, availability of water sources (in quality and quantity), and successive uses for applications requiring progressively lower water quality, water recycle and discharge standards.

Industrial water treatment facility consists of processes such as aeration, filtration, and clarification to remove materials from water such as suspended or dissolved solids, hardness and dissolved gases from water that may cause problems. After this basic treatment, the water is divided into different streams, some are used without further treatment and the rest are treated for specific applications like

- Addition of either hydrazine or sulfite to remove dissolved oxygen.
- Prevention of formation of calcium deposits by the addition of chelating agents to bind the dissolved calcium.
- Addition of precipitants such as phosphate for calcium removal.
- Treatment with dispersants to inhibit scale formation.
- Prevention of corrosion by the addition of inhibitors.
- pH adjustment.
- Disinfection for food processing uses or to prevent bacterial growth in cooling water

5.2. Mitigation Techniques for Wastewater Pollution

Following are the mitigation techniques for waste water pollution:

Wastewater treatment and safe disposal is a necessity as it will protect the environment and help in its conservation. Waste water treatment purpose is used to remove the contaminants from water so that the treated water meets the acceptable quality standards. The sewage treatment plants were constructed in end of nineteenth century and were designed to remove suspended matter alone by the principal of simple gravity settling. Later, these become apparent primary treatment alone as it was insufficient to protect the water quality of the receiving water body. It was due to the presence of organic material in colloidal and dissolved form in the sewage after settling. Thus, several treatment systems called secondary treatment were developed with the main objective of removal of organic material. In secondary treatments, biological methods are used. It was again found that the discharge of effluents from efficient secondary treatment plant can also lead to the deterioration of the quality of receiving water body due to discharge of ammonia in effluent. In water body, biological oxidation of ammonia to nitrate takes place known as nitrification which decreases oxygen in the water (Asolekar et al nd; Pokhrel, 2004).

Waste water treatment processes can be broadly classified as physical, chemical or biological. Physical processes comprises of screening, sedimentation, floatation and filtration. Chemical processes use chemical properties of the impurities of the added reagents. Commonly used chemical processes are precipitation, coagulation and disinfection. Biological processes use biochemical reactions; typical examples are bio-filtration and activated sludge process. These processes are grouped to form primary, secondary and tertiary treatment. The schematic diagram of waste water treatment is shown in Figure 9. In primary treatment suspended solids and floating material is removed. In secondary treatment or biological treatment, organic matter is removed. In tertiary treatment or advanced waste treatment physical, chemical or biological processes or their various combinations are used depending on the impurities to be removed like soluble non-biodegradable organic compounds like inorganic nutrients and salts, trace contaminants of various types, and dissolved inorganic salts. These methods are expensive and are used only when water produced is required of higher quality than that by secondary treatment.

1. **Primary Treatment:** It is sub-divided into following treatment methods.
2. **Pre-Treatment:** It consists of screening and grit removal. Screening removes or reduces the size of trash and large solids that get into sewage system. These solids are collected on screens and scraped off for subsequent disposal. After screening, the waste water is allowed to enter a grit chamber for the removal of

Figure 9. Schematic diagram of waste water treatment

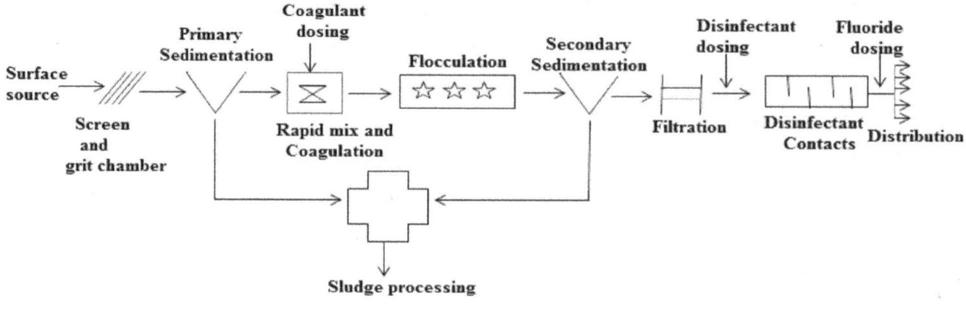

inorganic grit consisting of sand, gravels and pebbles. Grit is normally allowed to settle in a tank under conditions of low flow velocity and it is then scraped mechanically from the bottom of the tank.

3. **Primary Sedimentation:** It removes both the settleable and floatable solids. On addition of chemicals (lime and alum) flocculant particles will get settled. The floating material in the primary settling basin is grease. Grease floats on the surface which is removed by skimming. This process of flocculant settling takes place when the settling velocity of the particles increases due to coalescence with other particles in primary clarifiers. The particle removal efficiency depends on both the overflow and bed depth.
4. **Secondary or Biological Treatment:** Effluent from primary treatment still contains 40 to50% of original suspended concentration and virtually contain all dissolved organic and inorganic. Oxygen is supplied to the microorganisms and they take organic matter as food in controlled conditions so that most of the BOD is removed in the treatment plant rather than in the water course. The requirements of a biological waste treatment process are an adequate amount of bacteria, oxygen and some means of achieving contact between the bacteria and the organics. Biological waste treatment systems commonly used are suspended culture system and the attached culture system. In suspended culture, the microorganisms are suspended in wastewater as flocs which are surrounded by wastewater which contain their food and other essential elements. Attached culture system consist of masses of microorganisms adhere in an inert surface and wastewater contain organic matter passes over the microbial film. In both systems, organic matter is metabolized to stable inorganic forms. When handling large volume of waste, the suspended culture systems are more reliable. It is sub-divided into following treatment methods.
5. **Suspended Culture System:** Activated sludge process is a suspended culture system (Figure 10). In this process settled sludge containing microorganisms is returned to the reactor to increase the availability of microorganism biomass and this leads to the speeding up of the reactions. This process is aerobic in nature.

The essential feature of the process is an aeration tank where the organic matter is brought into intimate contact with the sludge from the secondary clarifier. This sludge is heavily laden with microorganisms which are in an active state of growth. Air is introduced into the tank, either in the form of bubbles through diffusers or by surface aerators. The microorganisms utilise oxygen in the air and convert the organic matter containing N and P into stabilized, low energy compounds such NO_3^-, SO_4^{2-}, NH_4^+, $H_2PO_4^-$, HPO_4^{2-}, CO_2, H_2O and new cells. The effluent from the aeration tank containing the flocculent biomass known as the sludge is separated in

Figure 10. Activated Sludge

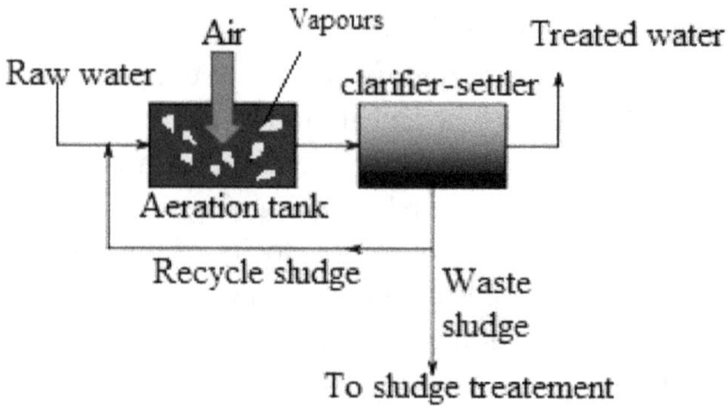

a settling tank sometimes called a secondary settler or clarifier. These solids settle out in the settler and a fraction of them is discarded. Part of the solids is recycled as return sludge to the head of the aeration tank and comes into contact with fresh sewage. The combination of high concentration of hungry cells in the return sludge and a rich food source in the influent sewage provides optimum conditions for waste degradation. In the activated sludge process, BOD is removed by two path ways. Organic matter is oxidised in the course of providing energy for the metabolic processes of the microorganism, and Synthesis and incorporation of organic matter into cell mass. In the first pathway, carbon is removed in the gaseous form as CO_2. The second pathway provides for removal of carbon as a solid in biomass. That portion of the carbon converted to CO_2 is vented to the atmosphere and does not present a disposal problem. What remains to be disposed of is a mixture of solids and water called sludge. The collection, processing and disposal of sludge can be the most costly and complex aspect of waste water treatment. The concentration of solids in the primary sewage sludge is about 5%; in the activated sludge it is less than 1% and the sludge from trickling filters has about 2% solids. This means that the sludge is composed almost entirely of water and volume reduction is the key to economic disposal. In addition to reducing its water content, the sludge must be stabilised so that its biological activity and tendency towards putrefaction are reduced drastically. Activated sludge treatment produces more microorganisms than necessary and if the microorganisms are not removed, their concentration will soon increase and clog the system with solids. Some of the microorganisms must therefore be wasted and the disposal of such waste activated sludge is one of the most difficult aspects of waste treatment.

The significant process that occurs during biological waste treatment is nitrification. In this process, ammonium ion is oxidised, under appropriate conditions, first to nitrite by Nitrosomonas bacteria.

$$2NH_4^+ + 3O_2 \rightarrow 4H^+ + NO_2^- + 2H_2O$$

$$2NO_2^- + O_2 \rightarrow 2NO_3^-$$

The above reactions occur in the aeration tank of the activated sludge plant and are favoured in general by long retention times, low organic loadings, large amounts of suspended solids and high temperatures.

Other suspended culture biological systems for treating wastewater are oxidation ponds, stabilization pond. Oxidation ponds (Figure 11) are large, shallow typically 1-2m deep, where raw or partially treated sewage is decomposed by microorganisms. The conditions are similar to those that prevail in a eutrophic lake. The ponds can be designed to maintain aerobic conditions throughout, but more often the decomposition taking place near the surface is aerobic, while that near the bottom is anaerobic. Such ponds having a mixture of aerobic and anaerobic conditions are called facultative ponds. The oxygen required for aerobic decomposition is derived from surface aeration and algal photosynthesis; deeper ponds called lagoons are mechanically aerated. Oxidation ponds can be designed to provide complete treatment to raw sewage but they require good deal of space. These ponds have been used extensively in small communities where land constraints are not so critical. They are easy to build and manage. They accommodate large fluctuations in flow, and they can provide treatment that approaches that of conventional biological systems but at much lower cost. The effluent however may contain undesirable concentrations of algae and, especially in the winter when less oxygen is liberated by photosynthesis which may produce unpleasant odors. They have the disadvantage that the effluent may not meet the EPA secondary treatment requirement of 30 mg/L BOD5 and suspended solids. However they are simple and effective in destroying pathogenic organisms which make these ponds useful in developing countries. Oxidation ponds are also used to supplement secondary treatment and in such cases they are called polishing ponds.

1. **Attached Culture System:** In attached culture system reactors wastewater is contacted with microbial films attached to surfaces. Surface areas of bifilm for growth of microorganism are increased by placing porous medium in reactor. When solid medium is used the reactor is known as trickling filter (Figure 12). When rotating disks are partially submerged in wastewater it is called rotating biological contractors. Conventional trickling filters normally consist of a rock bed, 1 to 3 meters in depth with enough opening between the rocks to allow air to circulate easily. The influent is sprinkled over a bed of packing which is

Figure 11. Oxidation pond

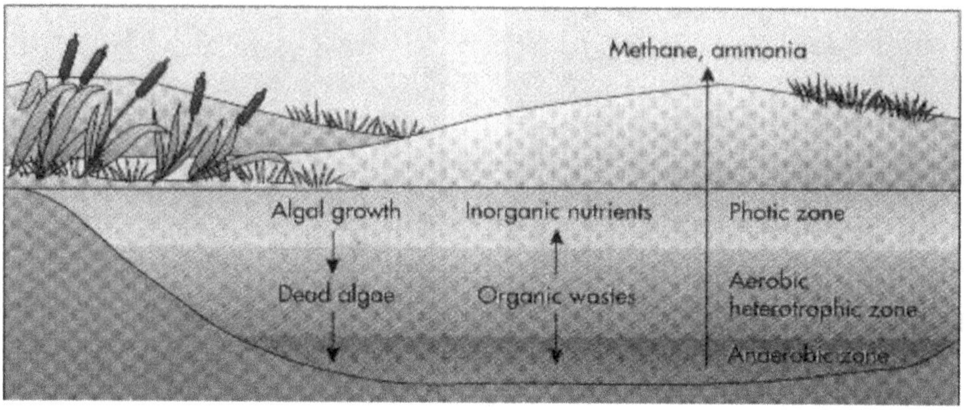

Figure 12. Trickling bed filter

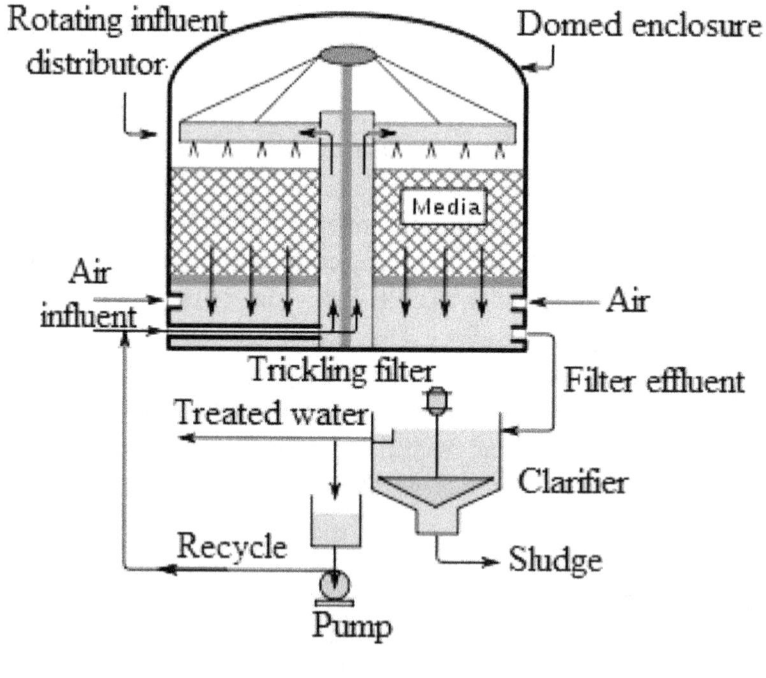

coated with a biological slime. As the liquid trickles over the packing, oxygen and the dissolved organic matter diffuse into the film to be metabolized by the microorganisms in the slime layer. End products such as CO_2, NO_3^- etc., diffuse back out of the film and appear in the filter effluent. Milk processing, paper mills and pharmaceuticals wastes are among those treated by trickling filters. Like all biological units trickling filters are affected by temperature; therefore cold weather slows down the biological activity in the filter.

2. **Rotating Biological Contactor:** A variation of this attached growth approach is provided by the rotating biological contactor (RBC) (Figure 13). An RBC consists of a series of closely spaced, circular, plastic disks, which are typically 3.6 m in diameter and attached to a rotating horizontal shaft. The bottom of 40% of each disc is submerged in a tank containing the waste water to be treated. The biomass film that grows on the surface of the disks moves into and out of the waste water as RBC rotates. While the microorganisms are submerged in waste water, they absorb organics; while they were rotated out of waste water, they are supplied with needed oxygen. By placing modular RBC units in series, treatment levels that exceed the conventional secondary treatment can be achieved. They are easier to operate under varying load conditions than trickling filters, since it is easier to keep the solid medium wet at all times.
3. **Tertiary Treatment:** Secondary treatment removes 80-90% of BOD and TSS and minor portions of nitrogen, phosphorus, and heavy metals. Tertiary treatment is also called as advanced treatment. The effluents from secondary sewage treatment plants contain both nitrogen (N) and phosphorus (P) as these are major ingredients of all fertilizers. When excess amounts of N and P are discharged, plant growth in the receiving waters may be accelerated which results in eutrophication in the water body receiving such waste. It is done to raise the effluent quality before it is discharged to environment such as river, lake, ground, etc., or to make it suitable for intended reuse. It includes nutrient

Figure 13. Rotating biological contactor

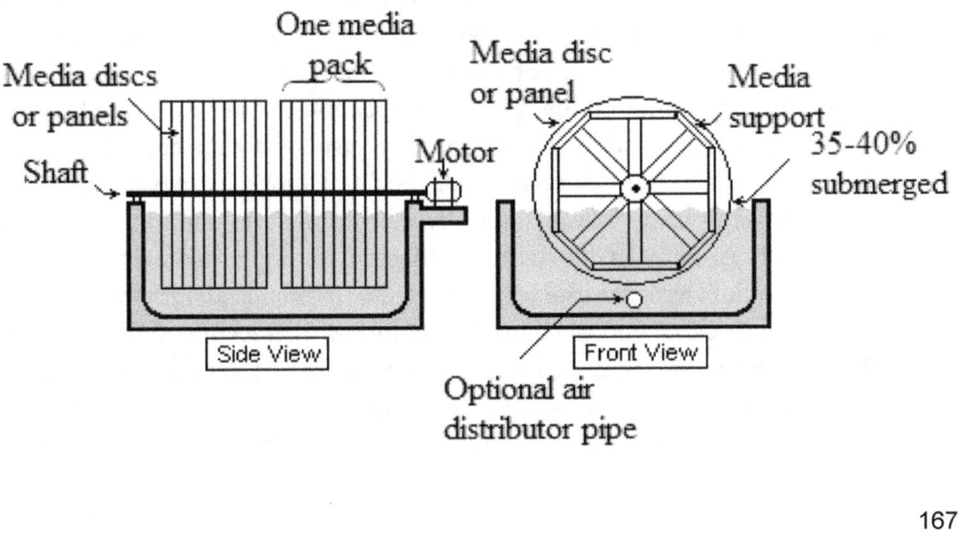

removal, membrane processes etc. Tertiary treatment is costly as compared to primary and secondary treatment methods. It is sub-divided into following treatment methods.

4. **Membrane Processes:** Membranes are made of semi-permeable materials designed to separate particulate, colloidal and dissolved substances from liquid solutes. They allow substances smaller than the membrane pores to flow through while holding back substances larger than the pores. Membranes are produced from a wide variety of materials such as cellulose acetate, polyamides, polysulfones, polypropylene, nylon, polyvinyl alcohol etc. It is further sub-divided into following remediation methods:
5. **Micro-Filtration (MF):** MF membranes (pores > 50 nm (nano-metre) are the least expensive membranes used in wastewater treatment for turbidity removal and solids separation after biological treatment as in Membrane Bioreactors (MBRs).
6. **Ultrafiltration (UF):** UF membranes (pore sizes 2-50 nm) have been used in wastewater treatment for many of the same applications as MF membranes except that UF systems give a better separation of finer colloids, bacteria, viruses, etc.
7. **Nanofiltration (NF):** In NF membranes, the pores should be less than 2 nm. The pressures vary between 520-1400 kPa and flux rates vary from 200-800 L/ m^2/d. It is used in water purification for potable purpose.
8. **Reverse Osmosis (RO):** Membranes have pores < 2nm and have the lowest molecular weight cut-off; high operating pressures of > 1400 kPa and Flux rates vary from 300 – 500 L/ m^2/d. RO is used in desalination operations. RO water is further treating after pre-treated by MF & UF to produce waters of high quality for indirect reuse applications.
9. **Chlorination:** Chlorine (Cl_2) is one of the most widely used disinfectants for water disinfection. It can be applied for the deactivation of most pathogenic microorganisms in drinking water, swimming pool water, wastewater, for textile bleaching, etc. and it is fairly persistent and relatively cheap. It is found in the form of the compounds on many different locations across the world because it is a very reactive and corrosive gas. It is usually found bond to sodium (Na) or in kitchen salt (sodium chloride; NaCl). Chlorine is mostly dissolved in seas and salty lakes. Large quantities of chlorine are found in the ground as rock salts or halite. Chlorine is available as compressed elemental gas, known as sodium hypochlorite solution (NaOCl) or solid calcium hypochlorite ($Ca(OCl)_2$). Factors which determine chlorine disinfection effectivity are chlorine concentrations, contact time, temperature, pH, number and types of microorganisms, concentrations of organic matter in the water. Types of chlorine used in disinfection are liquid chlorine (sodium hypochlorite) and dry chlorine

(calcium hypochlorite). Chlorination of water supply can be done by pump type (positive displacement) chlorinator and injector (aspirator) chlorinator.

10. **Biological Nutrient Removal (BNR):** It is a process used for nitrogen and phosphorus removal from wastewater before it is discharged into surface or ground water. The rising concentration of harmful nutrient compounds in municipal wastewater treatment plant discharge causes cultural eutrophication (nutrient enrichment due to human activities) in surface waters i.e. summer algal blooms. It can cause various problems to ecosystems i.e. fish kills, murky water, depletion of flora and fauna, etc. Wastewater treatment facilities are gradually being required to implement the processes that significantly reduce nutrient concentrations to safe levels. It usually involves major process modifications to wastewater treatment plant, such as making a portion of the aeration basin anaerobic and/or anoxic, which reduces the aerobic volume and limits nitrification capacity. The most effective solution for BNR is the active cell process developed by Headworks BIO used in Integrated Fixed-film Activated Sludge (IFAS) technology. Active cell is placed in activated sludge basins. IFAS technology is a hybrid process that describes attached-growth biological systems is combined with those of the suspended growth activated sludge process.

6. GROUNDWATER POLLUTION

Ground water contamination is generally irreversible i.e. once it is contaminated; it is difficult to restore the original water quality of the aquifer. The risk of groundwater pollution is increasing both from the disposal of water materials and from the widespread use by industry and agriculture of potentially polluting chemicals in the environment. Excessive mineralization of groundwater degrades its quality and produces an objectionable taste, odour and excessive hardness. Although the soil mantle through which water passes acts as an adsorbent retaining a large part of colloidal and soluble ions with its cation exchange capacity, but ground water is not completely free from the menace of chronic pollution. Therefore, it is always better to protect ground water in the first place rather than relying on technology to clean up contaminated water at a later stage.

Groundwater pollution can occur either as discrete, point sources (e.g. from landfills), or from the wider, more diffuse use of chemicals such as the application to and fertilizers and pesticides and the deposition of airborne pollutants in heavily industrialized regions (Singh et al, 1998). Changes in groundwater quality may result from direct or indirect anthropogenic activities. Direct influence occurs as a result of the introduction of natural or artificial substances derived from human activities into groundwater. Indirect influences are those changes in groundwater quality caused by human interference with hydrological, physical and biochemical processes, but without the addition of substances.

Pesticides used on farms or on lawns can cause serious ground water pollution. Improper pesticides use become also the cause of humans and animals death and also have adverse effect on aquatic life in the nearby streams. Septic systems can be a source of ground water pollution .If a septic system is not working properly it can contaminate groundwater with bacteria, viruses, household chemicals. Animal wastes are the source of bacteria and nitrates. Lagoons used to trap animal wastes can be a source of ground water pollution if they leak or if water table is very close to the land surface. Cadmium pollution can cause itai-itai disease, which affects lung and liver cancer. This is due to groundwater pollution.

The main factors which affecting ground water pollution are rain fall pattern, depth of water table, distance from the source of contamination, soil properties such as texture, structure and filtration rate.

There are various sources of groundwater pollution enlisted as follows:

1. **Natural Substances:** Some groundwater pollution occurs naturally. The toxic metal arsenic, for instance, is commonly found in the sediments or rock of the western United States and can be present in groundwater at concentrations that exceed safe levels for drinking water. Radon gas is a radioactive product of the decay of naturally occurring uranium in the earth's crust. Groundwater entering a house through a home water-supply system might release radon indoors where it could be breathed.
2. **Petroleum-Based Fuels:** One of the best known classes of groundwater contaminants includes petroleum-based fuels such as gasoline and diesel. Nationally, the U.S. Environmental Protection Agency (EPA) has recorded that there have been over 400,000 confirmed releases of petroleum-based fuels from leaking underground storage tanks. Gasoline consists of a mixture of various hydrocarbons (chemicals made up of carbon and hydrogen atoms) that evaporate easily, dissolve to some extent in water, and often are toxic. Benzene, a common component of gasoline, is considered to cause cancer in humans, whereas other gasoline components, such as toluene, ethylbenzene, and xylene, are not believed to cause.
3. **Chlorinated Solvents:** Another common class of groundwater contaminants includes chemicals known as chlorinated solvents. One example of a chlorinated solvent is dry-cleaning fluid also known as perchloroethylene. These chemicals are similar to petroleum hydrocarbons having carbon and hydrogen atoms, but also have chlorine atoms in their structure. As a general rule, the chlorine present in chlorinated solvents makes this class of compounds more toxic than

fuels. Unlike petroleum-based fuels, solvents are usually heavier than water and thus tend to sink to the bottoms of aquifers. This makes solvent-contaminated aquifers much more difficult to clean up than those contaminated by fuels.

6.1. Mitigation Measures for Ground Water Pollution

Groundwater is also known as sub surface water. It is the water present below ground surface and saturates the pore spaces in the sub-surface. Around 25 to 40% of water drawn from boreholes and wells is used for drinking purposes. Groundwater is also used by farmers for irrigation and by industrialist for manufacturing processes. Generally groundwater is very clean until it is been contaminated by the human activities or some time by natural conditions.

Groundwater gets polluted when pollutants released to field/ground make their way to groundwater and contaminates its quality. Contaminant plume is created in the aquifer (Figure 14). As the water moves and due to dispersion in the aquifer pollutant spreads on the wider area. This movement of plume along with the water can be studied through hydrological transport model. Groundwater pollution occurs from on-site sanitation systems, landfills, effluent from wastewater treatment plants,

Figure 14. Confined/ Unconfined Aquifers

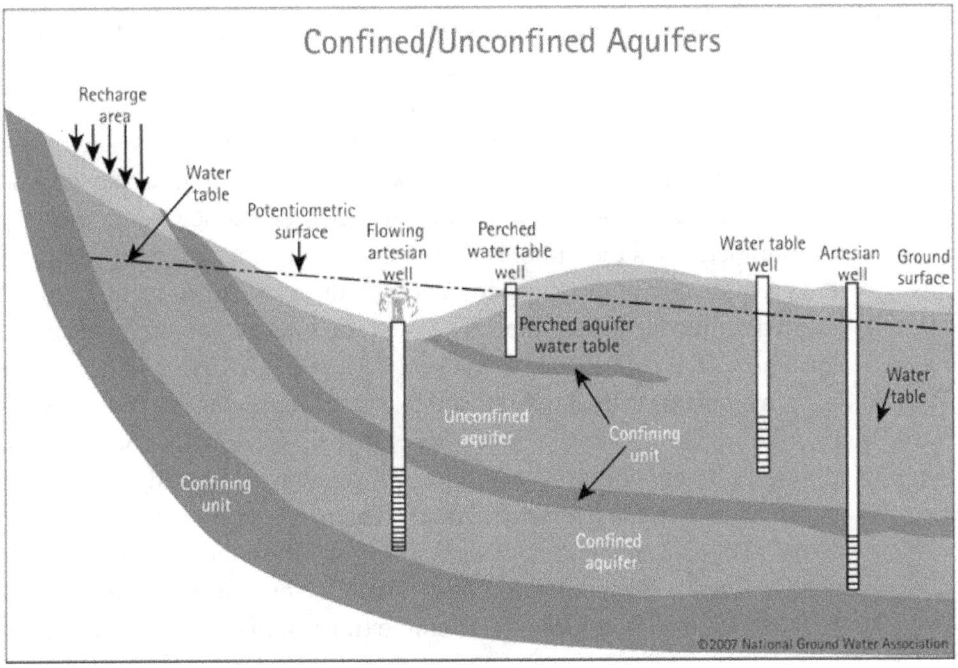

leaking sewers, petrol stations or from over application of fertilizers in agriculture. Groundwater contaminations also occur from naturally occurring contaminants, such as arsenic or fluoride (Kaluarachchi, 2001). The use of contaminated groundwater can lead to hazards to public health through poisoning or the spread of disease.

Detection of groundwater pollution is difficult than the surface water pollution, as groundwater moves large distance in aquifers. Groundwater pollution is detected when it is being noticed somewhere. Non-porous layer of clay in sub-surface partially purify water of bacteria by filtration method. Groundwater moving through open fractures and caverns is not filtered and can be transported as easily as surface water. In fact, this can be aggravated by the human tendency to use natural sinkholes as dumps.

Pollutants and contaminants can be removed from ground water by various techniques and making it safe for use. For the treatment of ground water, following treatment (or remediation) techniques (Bedient et al, 1999) are enlisted:

1. Physical treatment technologies
2. Chemical treatment technologies
3. Biological treatment technologies

1. **Physical Treatment Technologies:** This includes the following remediation techniques used:
    a. Pump and treat is one of the most widely used ground water remediation technologies in past. In this process, ground water is pumped to the surface and is coupled with either biological or chemical treatments to remove the impurities. Treated water is either used or returned to the aquifers. Moreover, the cost and time required increases with successive clean ups.
    b. Air sparging is the process of blowing air directly into the ground water (Figure 15). Addition of air enhances the biodegradation of the contaminants. As the bubbles rise, the contaminants are removed from the groundwater by physical contact with the air (i.e., stripping) and are carried up into the unsaturated zone (i.e., soil). As the contaminants move into the soil, a soil vapour extraction system is usually used to remove vapours. It helps in treat of immiscible, volatile contaminant source zones at or below the capillary fringe. Generally, air sparging is not useful for plume treatment or plume barriers when dealing with readily degradable compounds (by natural attenuation). This process is beneficial in the removal of contaminants such as benzene and gasoline. If air sparging is used without soil vapour extraction, it is called biosparging. Biosparging is applied only when location is remote (no risk to receptors); sufficient

Figure 15. Schematic diagram of an In situ air sparging system combined with soil vapour extraction

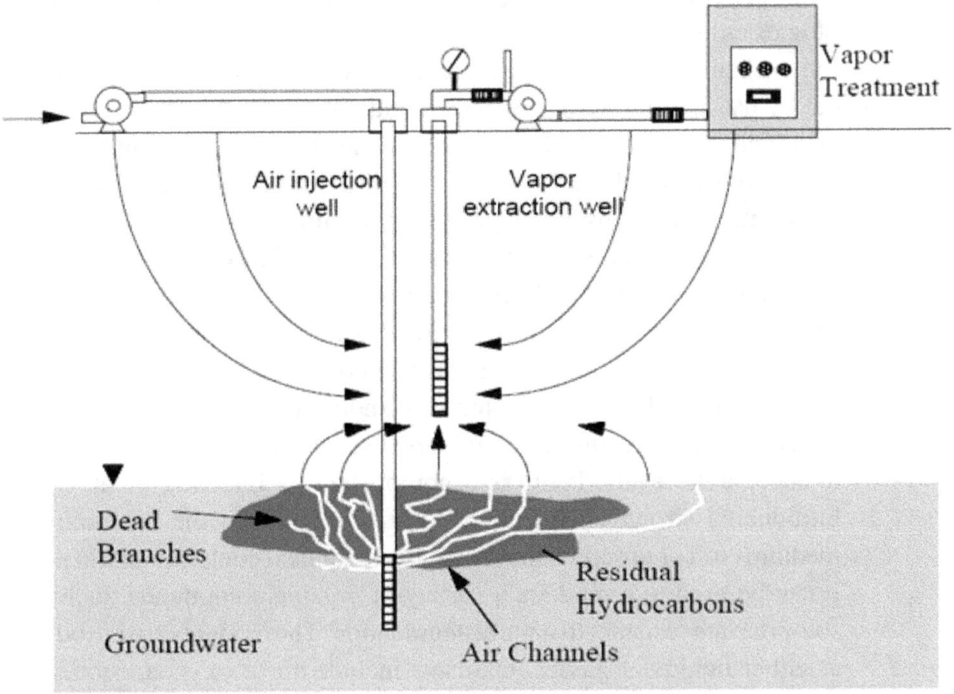

and rapid vapour phase biodegradation in vadose zone; low volatilization rate, and soil gas concentartion are below levels of concern for exposure.

c. Dual phase vacuum extraction (DPVE) (also known as multi-phase extraction) is a technology that uses a high-vacuum system to remove both contaminated groundwater and soil vapour. In DPVE systems, a high-vacuum extraction well is installed with its screened section in the zone of contaminated soils and groundwater. Fluid/vapour extraction systems depress the water table and water flows faster to the extraction well. DPVE removes contaminants from above and below the water table. As the water table around the well is lowered from pumping, unsaturated soil is exposed. This area, called the capillary fringe, is often highly contaminated, as it holds undissolved chemicals, chemicals that are lighter than water and vapors that have escaped from the dissolved groundwater below. Contaminants in the newly exposed zone can be removed by vapor extraction. Once above ground, the extracted vapors and liquid-phase organics and groundwater are separated and treated. Use of dual-phase vacuum extraction with these technologies can shorten the cleanup time at a site, because the capillary fringe is often the most contaminated area.

2. **Chemical Treatment Technologies:** Following are the chemical treatment technologies enlisted as:
    a. Ion exchange for ground water remediation is virtually always carried out by passing the water downward under pressure through a fixed bed of granular medium (either cation exchange media or anion exchange media) or spherical beads. Cations are displaced by certain cations from the solutions and ions are displaced by certain anions from the solution. Ion exchange media most often used for remediation are zeolites (both natural and synthetic) and synthetic resins.
    b. In situ chemical oxidation (ISCO) is a form of advanced oxidation processes and advanced oxidation technology. It is an environmental remediation technique used for soil and/or groundwater remediation to reduce the concentrations of targeted environmental contaminants to acceptable levels. ISCO is accomplished by injecting or otherwise introducing strong chemical oxidizers directly into the contaminated medium (soil or groundwater) to destroy chemical contaminants in place. It can be used to remediate a variety of organic compounds, including some that are resistant to natural degradation. The oxidants are introduced as either liquids or gasses. Oxidants include air or oxygen, ozone, and certain liquid chemicals such as hydrogen peroxide, permanganate and persulfate. Ozone and oxygen gas can be generated on site from air and electricity and directly injected into soil and groundwater contamination. The process has the potential to oxidize and/or enhance naturally occurring aerobic degradation. Chemical oxidation has proven to be an effective technique for dense non-aqueous phase liquid or DNAPL when it is present. The remediation of certain organic substances such as chlorinated solvents (trichloroethene and tetrachloroethene), and gasoline-related compounds (benzene, toluene and xylenes) by ISCO is possible. Some other contaminants can be made less toxic through chemical oxidation.
    c. Surfactant enhanced recovery increases the mobility and solubility of the contaminants absorbed to the saturated soil matrix or present as dense non-aqueous phase liquid. Surfactant-enhanced recovery injects surfactants (surface-active agents that are primary ingredient in soap and detergent) into contaminated groundwater. A typical system uses an extraction pump to remove groundwater downstream from the injection point. The extracted groundwater is treated aboveground to separate the injected surfactants from the contaminants and groundwater. Once the surfactants have separated from the groundwater they are re-used. The surfactants

used are non-toxic, food-grade, and biodegradable. Surfactant enhanced recovery is used most often when the groundwater is contaminated by dense non-aqueous phase liquids (DNAPLs). These dense compounds, such as trichloroethylene (TCE), sink in groundwater because they have a higher density than water. They then act as a continuous source for contaminant plumes that can stretch for miles within an aquifer. These compounds may biodegrade very slowly. They are commonly found in the vicinity of the original spill or leak where capillary forces have trapped them.

d. Permeable reactive barrier (PRB) (also known as permeable reactive treatment zone (PRTZ)) is the latest technique recognized as being a cost-effective technology for in situ (at the site) groundwater remediation (Figure 16). PRBs are barriers which allow some, but not all materials to pass through. It is also defined as in situ treatment zone that passively captures a plume of contaminants and removes or breaks down the contaminants, releasing uncontaminated water. Some permeable reactive barriers utilize chemical processes such as (1) sorption and precipitation; (2) chemical reactions; and (3) reactions involving biological mechanisms to achieve groundwater remediation. Advantages of PRB are: (i) typical treatment is passive; (ii) potentially lower operation and maintenance costs; (iii) allows full economic use of a property; and (iv) no above ground structures or routine day-to-day labour attention required.

Figure 16. In-situ permeable barrier

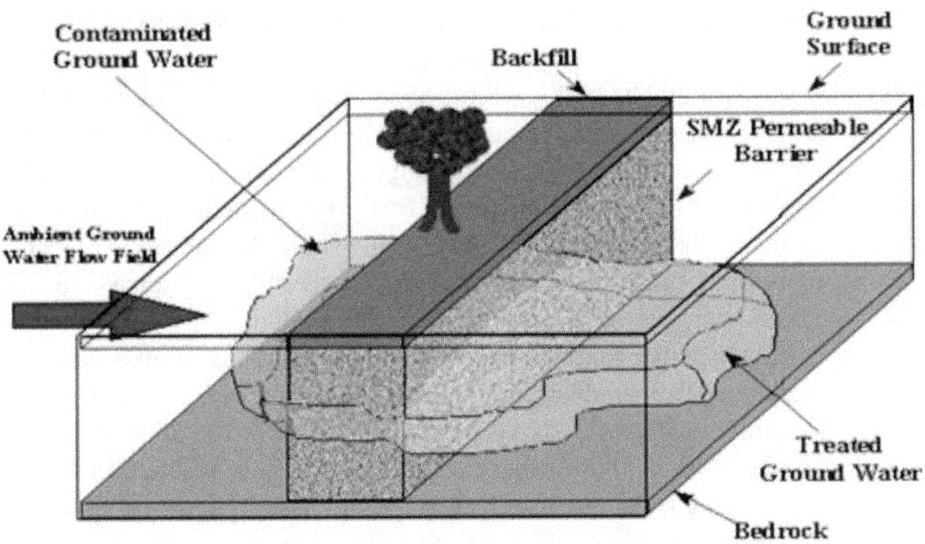

e. **Biological Treatment Technologies:** Following are the biological treatment technologies enlisted as:

3. **Biostimulation:** Treatment uses naturally occurring organisms to break down hazardous substances into less toxic or non-toxic substances. Technologies can be generally classified as in situ or ex situ. In situ bioremediation involves treating the contaminated material at the site, while ex situ involves the removal of the contaminated material to be treated elsewhere. Bioremediation may occur on its own (natural attenuation or intrinsic bioremediation) or may only effectively occur through the addition of fertilizers, oxygen, etc., that help in enhancing the growth of the pollution-eating microbes within the medium.
   a. **Bio Augmentation:** If a treatability study shows no degradation (or an extended lab period before significant degradation is achieved) in contamination contained in the groundwater, then inoculation with strains known to be capable of degrading the contaminants may be helpful. This process increases the reactive enzyme concentration within the bioremediation system and subsequently may increase contaminant degradation rates over the non-augmented rates.
   b. **Bioventing:** Bioventing is an in situ remediation technology that uses microorganisms to biodegrade organic constituents in the groundwater system. Bioventing enhances the activity of indigenous bacteria and archaea and stimulates the natural in situ biodegradation of hydrocarbons by inducing air or oxygen flow into the unsaturated zone and, if necessary, by adding nutrients. During bioventing, oxygen may be supplied through direct air injection into residual contamination in soil. Bioventing primarily assists in the degradation of adsorbed fuel residuals, but also assists in the degradation of volatile organic compounds (VOCs) as vapours move slowly through biologically active soil. \
   c. **Bioslurping:** The biological processes in the term bioslurping refer to aerobic biological degradation of the hydrocarbons when air is introduced into the unsaturated zone. Bioslurping combines elements of bioventing and vacuum-enhanced pumping of free-product that is lighter than water known as light non-aqueous phase liquid (LNAPL). It is used to recover free-product from the groundwater and soil and to bioremediate soils. The bioslurper system uses a slurp tube that extends into the free-product layer. Much like a straw in a glass draws liquid; the pump draws liquid (including free-product) and soil gas up the tube in the same process stream. Pumping lifts LNAPLs, such as oil, off the top of the water table and from the capillary fringe (i.e., an area just above the saturated zone, where water is held in place by capillary forces). The LNAPL is brought to the surface where it is separated from water and air.

d. **Phytoremediation:** In the phytoremediation process, certain plants and trees are planted, whose roots absorb contaminants from ground water over time, and are harvested and destroyed. This process can be carried out in areas where the roots can tap the ground water.

7. CONCLUSION

There are number of factors that cause water pollution leading to adverse health effects. The associated burden of diseases can be substantial which makes it important to develop various control strategies. This will significantly help in disease control so as to sustain better human life. Many countries across the world face major water quantity and quality challenges, compounded by the effects of rapid industrialization. Concerted actions are needed to safely manage the use of toxic chemicals and to develop monitoring and regulatory guidelines.

REFERENCES

Abdulraheem, M. Y. (1989). Health considerations in using treated industrial and municipal effluents for irrigation. *Desalination DSLNAH*, *72*(1-2), 81–113. doi:10.1016/0011-9164(89)80028-1

Aggarwal, S. K. (2005). *Water pollution*. Retrieved from https://books.google.co.in/books?id=VRl-24p3ju8C&printsec=frontcover#v=onepage&q&f=false

Asolekar & Arceivala. (n.d.). *Wastewater treatment for pollution control and reuse* (3rd ed.). Mc-Graw Hill Education.

Bedient, P. B., Rifai, H. S., & Newell, C. J. (1999). *Ground Water Contamination: Transport and Remediation* (2nd ed.). Prentice Hall Publisher. Retrieved from https://books.google.co.in/books/about/Ground_Water_Contamination.html?id=6RpSAAAAMAAJ&redir_esc=y

Ellis, K. V. (1989). *Surface water pollution and its control*. Retrieved from http://onlinelibrary.wiley.com/doi/10.1002/rrr.3450050408/abstract

Goel, P. K. (2006). *Water Pollution: Causes, Effects and Control*. Retrieved from https://books.google.co.in/books/about/Water_Pollution.html?id=4R9CYYoiFCcC

Halder, J., & Islam, N. (2015). J.N. Halder1, M.N. Islam, Water Pollution and its Impact on the Human Health. *Journal of Environment and Human*, *2*(1), 36–46. doi:10.15764/EH.2015.01005

Kaluarachchi, J. J. (2001). *Groundwater Contamination by Organic Pollutants*. Retrieved from http://ascelibrary.org/doi/book/10.1061/9780784405277

Masters, G. M., & Ela, W. P. (2007). *Introduction to Environmental Engineering and Science*. Retrieved from http://tradownload.co/results/introduction-to-environmental-engineering-and-sciences-by-gilbert-m-and-masters.html

Mukherjee & Nelliyat. (n.d*Groundwater Pollution and Emerging Environmental Challenges of Industrial Effluent Irrigation in Mettupalayam Taluk, Tamil Nadu, Comprehensive Assessment of Water Management in Agriculture*. International Water Management Institute.

Peavy, H. S., Rowe, D. R., & Tchobanoglous, G. (1985). *Environmental engineering* (International Editions). McGraw-Hill; http://trove.nla.gov.au/work/17834578

Pokhrel, D., & Viraraghavan, T. (2004). Treatment of pulp and paper mill wastewater—a review. *The Science of the Total Environment*, *333*(1-3), 37–58. doi:10.1016/j.scitotenv.2004.05.017 PMID:15364518

Rao, N. S., Gurunadha Rao, V. V. S., & Gupta, C. P. (1998). Groundwater pollution due to discharge of industrial effluents in Venkatapuram area, Visakhapatnam, Andhra Pradesh, India. *Environmental Geology, 33*(4), 289–294. doi:10.1007/s002540050248

Singh, K. P., & Parwana, H. K. (1998). Groundwater pollution due to industrial wastewater in Punjab state and strategies for its control. *Indian Journal of Environmental Protection, 19*, 241–244.

Subrahmanyam, K., & Yadaiah, P. (2001). Assessment of the impact of industrial effluents on water quality in Patancheru and environs, Medak district, Andhra Pradesh, India. *Hydrogeology Journal, 9*(3), 297–312. doi:10.1007/s100400000120

Conclusion

Modernization and intensive business in rapidly developing industrial enterprises brought with it a number of problems. Waste with significant economic potential is a major problem for the environment and increasing population. If necessary precautions are not taken, waste generated by businesses may pollute underground and surface water resources as potential pollutants. Therefore, storage and forecasting criteria must be examined to ensure that the waste generated by enterprises does not create adverse environmental conditions. In this book, the environmental impacts of pollution created in industrial enterprises and natural habitats are evaluated, and the negativities created by the waste and methods of removing it from the environment without causing environmental pollution examined.

Today, industrialization is important in terms of environmental pollution, and remains a key issue in evaluating and removing waste. As an aspect of waste management, the pollution problem must be considered alongside health protection and pollution control programs for the undertaking, transport, storage, and use of waste.

Waste repositories are important in industrial enterprises for the protection of underground and surface water resources and control of unwanted odor and pollution. If businesses want to generate income from waste, it must be stored in a suitable environment to prevent health problems. In the majority of enterprises surveyed, only the planning and construction of waste depots were considered, and auxiliary compartments, especially those storing solid and liquid waste, neglected. For most of the enterprises, there were no waste deposits, and the waste was poured in bulk in the open. The random storage of waste negatively affects human health. Furthermore, waste management is failing, because the desired quality cannot be obtained. This, along with creating unwanted health problems in the stables, reduces business profitability. Waste stored at random releases bad odors into surrounding areas, causing air pollution and laying the ground for the deterioration of environmental health and spread of various infectious diseases.

As a commercial substance, waste is used to increase crop production and in energy production. Used in this way, the soil balance is preserved in terms of organic matter. The amount of production, product quality, soil structure, and underground and surface waters begin to be negatively affected when the use of waste is measured to improve the physical structure of the soil and support energy production. Priorities do not focus on the problems caused by the waste used in the land. However, its rapid development in recent years has made waste a problem. Important here is that all types of waste in filtrate the permeable soil of contaminated water and reach the bottom water. This mixing is also the case for sloped land and surface waters in rainy weather. In addition, rivers can carry pollution to other regions, affecting the base water in that region. As contaminated ground waters are particularly rich in nitrates, surface waters start to reach high values in terms of the phosphorus and nitrogen content. One cause of fish deaths is the reduction of oxygen in the water, which is caused by the oxygen used during the decay of organic substances that come into contact with water. On the other hand, waste leaching into the surface waters accelerates the proliferation of algae and grasses in the environment, which then rotates while using oxygen, causing additional contamination.

The storage and management of waste water can be achieved by collecting it in pools. The investigations indicated that the use of plants to meet irrigation water and plant nutrient requirements through proper storage and management of waste water in the pools does not adversely affect plant quality and production. A higher volume of the product can obtained from contained plant nutrients using irrigation water. Important in practice is that the use of wastewater in agricultural production, which the plant needs, should not cause environmental pollution. Nitrogen and phosphorus are two factors affecting the amount of application of waste water. At high concentrations, wastewater containing nitrogen and phosphorus should be mixed with other usable water resources. In addition, since the waste water collected in the pools differs in terms of the concentration of the nutrient, the rate of application should be determined by analyzing the water before it is used to meet the need for irrigation water and nutrients.

The most effective way to remove waste and use it in crop production is using agricultural waste for agricultural purposes in accordance with the area and current techniques. Waste that causes environmental problems in industrial enterprises also has economic potential. Most waste is likely to be reused and recycled, which reduces environmental pollution through the evaluation of waste and its potential as an economic resource. In recent years, opportunities have been created that ensure less harm to the environment. Furthermore, processes such as ventilation, biogas production, composting, and drying have increased the effectiveness of waste as organic matter. In addition, odors emitted from the environment are reduced, disease agents killed, the level of waste matter and hydrogen content maintained or reduced, and the volume of emission gases significantly reduced. Dust and odorous emissions should be carefully managed in operating conditions. For example, waste gases should be discharged through chimneys into the atmosphere, that suitable for use should be stored, biogas facilities should be investigated, liquid waste collected and stored appropriately, and waste storage areas designed to ensure a storage capacity of three months' worth of waste. In this way, waste pollution in settlements or nearby areas can be prevented. Properly designed waste water ponds can help prevent environmental pollution by making it possible to use the accumulated wastewater throughout the year in the appropriate place and time while providing a healthier environment for people and other living things in the environment. While this type of system brings an additional cost to operations, the planned and proper use of waste water during the irrigation season can lower the cost of the operator's irrigation water. In the past, there have been legal regulations regarding the storage, management, and evaluation of waste in contemporary society, where environmental protection is a priority, as well as for the storage and management of waste water. Today, many countries have prohibited pollution of their underground and surface water resources through regulations. As such, the import, transportation, storage, and use of chemical substances, which are permanent in the air, water, or soil and disrupt the ecological balance, should be considered.

If the methods of this book on the protection of the environment are evaluated in this regard, it is possible that the negative impacts of industrial facilities on the environment can be mitigated.

Related References

To continue our tradition of advancing academic research, we have compiled a list of recommended IGI Global readings. These references will provide additional information and guidance to further enrich your knowledge and assist you with your own research and future publications.

Adeyemo, O. (2013). The nationwide health information network: A biometric approach to prevent medical identity theft. In *User-driven healthcare: Concepts, methodologies, tools, and applications* (pp. 1636–1649). Hershey, PA: IGI Global. doi:10.4018/978-1-4666-2770-3.ch081

Adler, M., & Henman, P. (2009). Justice beyond the courts: The implications of computerisation for procedural justice in social security. In A. Martínez & P. Abat (Eds.), *E-justice: Using information communication technologies in the court system* (pp. 65–86). Hershey, PA: IGI Global. doi:10.4018/978-1-59904-998-4.ch005

Aflalo, E., & Gabay, E. (2013). An information system for coping with student dropout. In L. Tomei (Ed.), *Learning tools and teaching approaches through ICT advancements* (pp. 176–187). Hershey, PA: IGI Global. doi:10.4018/978-1-4666-2017-9.ch016

Ahmed, M. A., Janssen, M., & van den Hoven, J. (2012). Value sensitive transfer (VST) of systems among countries: Towards a framework. *International Journal of Electronic Government Research*, *8*(1), 26–42. doi:10.4018/jegr.2012010102

Aikins, S. K. (2008). Issues and trends in internet-based citizen participation. In G. Garson & M. Khosrow-Pour (Eds.), *Handbook of research on public information technology* (pp. 31–40). Hershey, PA: IGI Global. doi:10.4018/978-1-59904-857-4.ch004

Aikins, S. K. (2009). A comparative study of municipal adoption of internet-based citizen participation. In C. Reddick (Ed.), *Handbook of research on strategies for local e-government adoption and implementation: Comparative studies* (pp. 206–230). Hershey, PA: IGI Global. doi:10.4018/978-1-60566-282-4.ch011

Aikins, S. K. (2012). Improving e-government project management: Best practices and critical success factors. In *Digital democracy: Concepts, methodologies, tools, and applications* (pp. 1314–1332). Hershey, PA: IGI Global. doi:10.4018/978-1-4666-1740-7.ch065

Akabawi, M. S. (2011). Ghabbour group ERP deployment: Learning from past technology failures. In E. Business Research and Case Center (Ed.), Cases on business and management in the MENA region: New trends and opportunities (pp. 177-203). Hershey, PA: IGI Global. doi:10.4018/978-1-60960-583-4.ch012

Akabawi, M. S. (2013). Ghabbour group ERP deployment: Learning from past technology failures. In *Industrial engineering: Concepts, methodologies, tools, and applications* (pp. 933–958). Hershey, PA: IGI Global. doi:10.4018/978-1-4666-1945-6.ch051

Akbulut, A. Y., & Motwani, J. (2008). Integration and information sharing in e-government. In G. Putnik & M. Cruz-Cunha (Eds.), *Encyclopedia of networked and virtual organizations* (pp. 729–734). Hershey, PA: IGI Global. doi:10.4018/978-1-59904-885-7.ch096

Akers, E. J. (2008). Technology diffusion in public administration. In G. Garson & M. Khosrow-Pour (Eds.), *Handbook of research on public information technology* (pp. 339–348). Hershey, PA: IGI Global. doi:10.4018/978-1-59904-857-4.ch033

Al-Shafi, S. (2008). Free wireless internet park services: An investigation of technology adoption in Qatar from a citizens' perspective. *Journal of Cases on Information Technology, 10*(3), 21–34. doi:10.4018/jcit.2008070103

Al-Shafi, S., & Weerakkody, V. (2009). Implementing free wi-fi in public parks: An empirical study in Qatar. *International Journal of Electronic Government Research, 5*(3), 21–35. doi:10.4018/jegr.2009070102

Aladwani, A. M. (2002). Organizational actions, computer attitudes and end-user satisfaction in public organizations: An empirical study. In C. Snodgrass & E. Szewczak (Eds.), *Human factors in information systems* (pp. 153–168). Hershey, PA: IGI Global. doi:10.4018/978-1-931777-10-0.ch012

Aladwani, A. M. (2002). Organizational actions, computer attitudes, and end-user satisfaction in public organizations: An empirical study. *Journal of Organizational and End User Computing*, *14*(1), 42–49. doi:10.4018/joeuc.2002010104

Allen, B., Juillet, L., Paquet, G., & Roy, J. (2005). E-government and private-public partnerships: Relational challenges and strategic directions. In M. Khosrow-Pour (Ed.), *Practicing e-government: A global perspective* (pp. 364–382). Hershey, PA: IGI Global. doi:10.4018/978-1-59140-637-2.ch016

Alshawaf, A., & Knalil, O. E. (2008). IS success factors and IS organizational impact: Does ownership type matter in Kuwait? *International Journal of Enterprise Information Systems*, *4*(2), 13–33. doi:10.4018/jeis.2008040102

Ambali, A. R. (2009). Digital divide and its implication on Malaysian e-government: Policy initiatives. In H. Rahman (Ed.), *Social and political implications of data mining: Knowledge management in e-government* (pp. 267–287). Hershey, PA: IGI Global. doi:10.4018/978-1-60566-230-5.ch016

Amoretti, F. (2007). Digital international governance. In A. Anttiroiko & M. Malkia (Eds.), *Encyclopedia of digital government* (pp. 365–370). Hershey, PA: IGI Global. doi:10.4018/978-1-59140-789-8.ch056

Amoretti, F. (2008). Digital international governance. In A. Anttiroiko (Ed.), *Electronic government: Concepts, methodologies, tools, and applications* (pp. 688–696). Hershey, PA: IGI Global. doi:10.4018/978-1-59904-947-2.ch058

Amoretti, F. (2008). E-government at supranational level in the European Union. In A. Anttiroiko (Ed.), *Electronic government: Concepts, methodologies, tools, and applications* (pp. 1047–1055). Hershey, PA: IGI Global. doi:10.4018/978-1-59904-947-2.ch079

Amoretti, F. (2008). E-government regimes. In A. Anttiroiko (Ed.), *Electronic government: Concepts, methodologies, tools, and applications* (pp. 3846–3856). Hershey, PA: IGI Global. doi:10.4018/978-1-59904-947-2.ch280

Amoretti, F. (2009). Electronic constitution: A Braudelian perspective. In F. Amoretti (Ed.), *Electronic constitution: Social, cultural, and political implications* (pp. 1–19). Hershey, PA: IGI Global. doi:10.4018/978-1-60566-254-1.ch001

Amoretti, F., & Musella, F. (2009). Institutional isomorphism and new technologies. In M. Khosrow-Pour (Ed.), *Encyclopedia of information science and technology* (2nd ed.; pp. 2066–2071). Hershey, PA: IGI Global. doi:10.4018/978-1-60566-026-4.ch325

Andersen, K. V., & Henriksen, H. Z. (2007). E-government research: Capabilities, interaction, orientation, and values. In D. Norris (Ed.), *Current issues and trends in e-government research* (pp. 269–288). Hershey, PA: IGI Global. doi:10.4018/978-1-59904-283-1.ch013

Anderson, K. V., & Henriksen, H. Z. (2005). The first leg of e-government research: Domains and application areas 1998-2003. *International Journal of Electronic Government Research, 1*(4), 26–44. doi:10.4018/jegr.2005100102

Anttiroiko, A. (2009). Democratic e-governance. In M. Khosrow-Pour (Ed.), *Encyclopedia of information science and technology* (2nd ed.; pp. 990–995). Hershey, PA: IGI Global. doi:10.4018/978-1-60566-026-4.ch158

Association, I. R. (2010). *Networking and telecommunications: Concepts, methodologies, tools and applications* (Vols. 1–3). Hershey, PA: IGI Global. doi:10.4018/978-1-60566-986-1

Association, I. R. (2010). *Web-based education: Concepts, methodologies, tools and applications* (Vols. 1–3). Hershey, PA: IGI Global. doi:10.4018/978-1-61520-963-7

Baker, P. M., Bell, A., & Moon, N. W. (2009). Accessibility issues in municipal wireless networks. In C. Reddick (Ed.), *Handbook of research on strategies for local e-government adoption and implementation: Comparative studies* (pp. 569–588). Hershey, PA: IGI Global. doi:10.4018/978-1-60566-282-4.ch030

Becker, S. A., Keimer, R., & Muth, T. (2010). A case on university and community collaboration: The sci-tech entrepreneurial training services (ETS) program. In S. Becker & R. Niebuhr (Eds.), *Cases on technology innovation: Entrepreneurial successes and pitfalls* (pp. 68–90). Hershey, PA: IGI Global. doi:10.4018/978-1-61520-609-4.ch003

Becker, S. A., Keimer, R., & Muth, T. (2012). A case on university and community collaboration: The sci-tech entrepreneurial training services (ETS) program. In Regional development: Concepts, methodologies, tools, and applications (pp. 947-969). Hershey, PA: IGI Global. doi:10.4018/978-1-4666-0882-5.ch507

Bernardi, R. (2012). Information technology and resistance to public sector reforms: A case study in Kenya. In T. Papadopoulos & P. Kanellis (Eds.), *Public sector reform using information technologies: Transforming policy into practice* (pp. 59–78). Hershey, PA: IGI Global. doi:10.4018/978-1-60960-839-2.ch004

Bernardi, R. (2013). Information technology and resistance to public sector reforms: A case study in Kenya. In *User-driven healthcare: Concepts, methodologies, tools, and applications* (pp. 14–33). Hershey, PA: IGI Global. doi:10.4018/978-1-4666-2770-3.ch002

Bolívar, M. P., Pérez, M. D., & Hernández, A. M. (2012). Municipal e-government services in emerging economies: The Latin-American and Caribbean experiences. In Y. Chen & P. Chu (Eds.), *Electronic governance and cross-boundary collaboration: Innovations and advancing tools* (pp. 198–226). Hershey, PA: IGI Global. doi:10.4018/978-1-60960-753-1.ch011

Borycki, E. M., & Kushniruk, A. W. (2010). Use of clinical simulations to evaluate the impact of health information systems and ubiquitous computing devices upon health professional work. In S. Mohammed & J. Fiaidhi (Eds.), *Ubiquitous health and medical informatics: The ubiquity 2.0 trend and beyond* (pp. 552–573). Hershey, PA: IGI Global. doi:10.4018/978-1-61520-777-0.ch026

Borycki, E. M., & Kushniruk, A. W. (2011). Use of clinical simulations to evaluate the impact of health information systems and ubiquitous computing devices upon health professional work. In *Clinical technologies: Concepts, methodologies, tools and applications* (pp. 532–553). Hershey, PA: IGI Global. doi:10.4018/978-1-60960-561-2.ch220

Buchan, J. (2011). Developing a dynamic and responsive online learning environment: A case study of a large Australian university. In B. Czerkawski (Ed.), *Free and open source software for e-learning: Issues, successes and challenges* (pp. 92–109). Hershey, PA: IGI Global. doi:10.4018/978-1-61520-917-0.ch006

Buenger, A. W. (2008). Digital convergence and cybersecurity policy. In G. Garson & M. Khosrow-Pour (Eds.), *Handbook of research on public information technology* (pp. 395–405). Hershey, PA: IGI Global. doi:10.4018/978-1-59904-857-4.ch038

Burn, J. M., & Loch, K. D. (2002). The societal impact of world wide web - Key challenges for the 21st century. In A. Salehnia (Ed.), *Ethical issues of information systems* (pp. 88–106). Hershey, PA: IGI Global. doi:10.4018/978-1-931777-15-5.ch007

Burn, J. M., & Loch, K. D. (2003). The societal impact of the world wide web-Key challenges for the 21st century. In M. Khosrow-Pour (Ed.), *Advanced topics in information resources management* (Vol. 2, pp. 32–51). Hershey, PA: IGI Global. doi:10.4018/978-1-59140-062-2.ch002

Bwalya, K. J., Du Plessis, T., & Rensleigh, C. (2012). The "quicksilver initiatives" as a framework for e-government strategy design in developing economies. In K. Bwalya & S. Zulu (Eds.), *Handbook of research on e-government in emerging economies: Adoption, e-participation, and legal frameworks* (pp. 605–623). Hershey, PA: IGI Global. doi:10.4018/978-1-4666-0324-0.ch031

Cabotaje, C. E., & Alampay, E. A. (2013). Social media and citizen engagement: Two cases from the Philippines. In S. Saeed & C. Reddick (Eds.), *Human-centered system design for electronic governance* (pp. 225–238). Hershey, PA: IGI Global. doi:10.4018/978-1-4666-3640-8.ch013

Camillo, A., Di Pietro, L., Di Virgilio, F., & Franco, M. (2013). Work-groups conflict at PetroTech-Italy, S.R.L.: The influence of culture on conflict dynamics. In B. Christiansen, E. Turkina, & N. Williams (Eds.), *Cultural and technological influences on global business* (pp. 272–289). Hershey, PA: IGI Global. doi:10.4018/978-1-4666-3966-9.ch015

Capra, E., Francalanci, C., & Marinoni, C. (2008). Soft success factors for m-government. In A. Anttiroiko (Ed.), *Electronic government: Concepts, methodologies, tools, and applications* (pp. 1213–1233). Hershey, PA: IGI Global. doi:10.4018/978-1-59904-947-2.ch089

Cartelli, A. (2009). The implementation of practices with ICT as a new teaching-learning paradigm. In A. Cartelli & M. Palma (Eds.), *Encyclopedia of information communication technology* (pp. 413–417). Hershey, PA: IGI Global. doi:10.4018/978-1-59904-845-1.ch055

Charalabidis, Y., Lampathaki, F., & Askounis, D. (2010). Investigating the landscape in national interoperability frameworks. *International Journal of E-Services and Mobile Applications*, *2*(4), 28–41. doi:10.4018/jesma.2010100103

Charalabidis, Y., Lampathaki, F., & Askounis, D. (2012). Investigating the landscape in national interoperability frameworks. In A. Scupola (Ed.), *Innovative mobile platform developments for electronic services design and delivery* (pp. 218–231). Hershey, PA: IGI Global. doi:10.4018/978-1-4666-1568-7.ch013

Chen, I. (2005). Distance education associations. In C. Howard, J. Boettcher, L. Justice, K. Schenk, P. Rogers, & G. Berg (Eds.), *Encyclopedia of distance learning* (pp. 599–612). Hershey, PA: IGI Global. doi:10.4018/978-1-59140-555-9.ch087

Chen, I. (2008). Distance education associations. In L. Tomei (Ed.), *Online and distance learning: Concepts, methodologies, tools, and applications* (pp. 562–579). Hershey, PA: IGI Global. doi:10.4018/978-1-59904-935-9.ch048

Chen, Y. (2008). Managing IT outsourcing for digital government. In A. Anttiroiko (Ed.), *Electronic government: Concepts, methodologies, tools, and applications* (pp. 3107–3114). Hershey, PA: IGI Global. doi:10.4018/978-1-59904-947-2.ch229

Chen, Y., & Dimitrova, D. V. (2006). Electronic government and online engagement: Citizen interaction with government via web portals. *International Journal of Electronic Government Research*, *2*(1), 54–76. doi:10.4018/jegr.2006010104

Chen, Y., & Knepper, R. (2005). Digital government development strategies: Lessons for policy makers from a comparative perspective. In W. Huang, K. Siau, & K. Wei (Eds.), *Electronic government strategies and implementation* (pp. 394–420). Hershey, PA: IGI Global. doi:10.4018/978-1-59140-348-7.ch017

Chen, Y., & Knepper, R. (2008). Digital government development strategies: Lessons for policy makers from a comparative perspective. In H. Rahman (Ed.), *Developing successful ICT strategies: Competitive advantages in a global knowledge-driven society* (pp. 334–356). Hershey, PA: IGI Global. doi:10.4018/978-1-59904-654-9.ch017

Cherian, E. J., & Ryan, T. W. (2014). Incongruent needs: Why differences in the iron-triangle of priorities make health information technology adoption and use difficult. In C. El Morr (Ed.), *Research perspectives on the role of informatics in health policy and management* (pp. 209–221). Hershey, PA: IGI Global. doi:10.4018/978-1-4666-4321-5.ch012

Cho, H. J., & Hwang, S. (2010). Government 2.0 in Korea: Focusing on e-participation services. In C. Reddick (Ed.), *Politics, democracy and e-government: Participation and service delivery* (pp. 94–114). Hershey, PA: IGI Global. doi:10.4018/978-1-61520-933-0.ch006

Chorus, C., & Timmermans, H. (2010). Ubiquitous travel environments and travel control strategies: Prospects and challenges. In M. Wachowicz (Ed.), *Movement-aware applications for sustainable mobility: Technologies and approaches* (pp. 30–51). Hershey, PA: IGI Global. doi:10.4018/978-1-61520-769-5.ch003

Chuanshen, R. (2007). E-government construction and China's administrative litigation act. In A. Anttiroiko & M. Malkia (Eds.), *Encyclopedia of digital government* (pp. 507–510). Hershey, PA: IGI Global. doi:10.4018/978-1-59140-789-8.ch077

Ciaghi, A., & Villafiorita, A. (2012). Law modeling and BPR for public administration improvement. In K. Bwalya & S. Zulu (Eds.), *Handbook of research on e-government in emerging economies: Adoption, e-participation, and legal frameworks* (pp. 391–410). Hershey, PA: IGI Global. doi:10.4018/978-1-4666-0324-0.ch019

Ciaramitaro, B. L., & Skrocki, M. (2012). mHealth: Mobile healthcare. In B. Ciaramitaro (Ed.), Mobile technology consumption: Opportunities and challenges (pp. 99-109). Hershey, PA: IGI Global. doi:10.4018/978-1-61350-150-4.ch007

Comite, U. (2012). Innovative processes and managerial effectiveness of e-procurement in healthcare. In A. Manoharan & M. Holzer (Eds.), *Active citizen participation in e-government: A global perspective* (pp. 206–229). Hershey, PA: IGI Global. doi:10.4018/978-1-4666-0116-1.ch011

Cordella, A. (2013). E-government success: How to account for ICT, administrative rationalization, and institutional change. In J. Gil-Garcia (Ed.), *E-government success factors and measures: Theories, concepts, and methodologies* (pp. 40–51). Hershey, PA: IGI Global. doi:10.4018/978-1-4666-4058-0.ch003

Cropf, R. A. (2009). ICT and e-democracy. In M. Khosrow-Pour (Ed.), *Encyclopedia of information science and technology* (2nd ed.; pp. 1789–1793). Hershey, PA: IGI Global. doi:10.4018/978-1-60566-026-4.ch281

Cropf, R. A. (2009). The virtual public sphere. In M. Pagani (Ed.), *Encyclopedia of multimedia technology and networking* (2nd ed.; pp. 1525–1530). Hershey, PA: IGI Global. doi:10.4018/978-1-60566-014-1.ch206

D'Abundo, M. L. (2013). Electronic health record implementation in the United States healthcare industry: Making the process of change manageable. In V. Wang (Ed.), *Handbook of research on technologies for improving the 21st century workforce: Tools for lifelong learning* (pp. 272–286). Hershey, PA: IGI Global. doi:10.4018/978-1-4666-2181-7.ch018

Damurski, L. (2012). E-participation in urban planning: Online tools for citizen engagement in Poland and in Germany. *International Journal of E-Planning Research*, *1*(3), 40–67. doi:10.4018/ijepr.2012070103 de Almeida, M. O. (2007). E-government strategy in Brazil: Increasing transparency and efficiency through e-government procurement. In M. Gascó-Hernandez (Ed.), *Latin America online: Cases, successes and pitfalls* (pp. 34–82). Hershey, PA: IGI Global. doi:10.4018/978-1-59140-974-8.ch002

de Juana Espinosa, S. (2008). Empirical study of the municipalitites' motivations for adopting online presence. In A. Anttiroiko (Ed.), *Electronic government: Concepts, methodologies, tools, and applications* (pp. 3593–3608). Hershey, PA: IGI Global. doi:10.4018/978-1-59904-947-2.ch262

de Souza Dias, D. (2002). Motivation for using information technology. In C. Snodgrass & E. Szewczak (Eds.), *Human factors in information systems* (pp. 55–60). Hershey, PA: IGI Global. doi:10.4018/978-1-931777-10-0.ch005

Demediuk, P. (2006). Government procurement ICT's impact on the sustainability of SMEs and regional communities. In S. Marshall, W. Taylor, & X. Yu (Eds.), *Encyclopedia of developing regional communities with information and communication technology* (pp. 321–324). Hershey, PA: IGI Global. doi:10.4018/978-1-59140-575-7.ch056

Devonshire, E., Forsyth, H., Reid, S., & Simpson, J. M. (2013). The challenges and opportunities of online postgraduate coursework programs in a traditional university context. In B. Tynan, J. Willems, & R. James (Eds.), *Outlooks and opportunities in blended and distance learning* (pp. 353–368). Hershey, PA: IGI Global. doi:10.4018/978-1-4666-4205-8.ch026

Di Cerbo, F., Scotto, M., Sillitti, A., Succi, G., & Vernazza, T. (2007). Toward a GNU/Linux distribution for corporate environments. In S. Sowe, I. Stamelos, & I. Samoladas (Eds.), *Emerging free and open source software practices* (pp. 215–236). Hershey, PA: IGI Global. doi:10.4018/978-1-59904-210-7.ch010

Diesner, J., & Carley, K. M. (2005). Revealing social structure from texts: Meta-matrix text analysis as a novel method for network text analysis. In V. Narayanan & D. Armstrong (Eds.), *Causal mapping for research in information technology* (pp. 81–108). Hershey, PA: IGI Global. doi:10.4018/978-1-59140-396-8.ch004

Dologite, D. G., Mockler, R. J., Bai, Q., & Viszhanyo, P. F. (2006). IS change agents in practice in a US-Chinese joint venture. In M. Hunter & F. Tan (Eds.), *Advanced topics in global information management* (Vol. 5, pp. 331–352). Hershey, PA: IGI Global. doi:10.4018/978-1-59140-923-6.ch015

Drnevich, P., Brush, T. H., & Luckock, G. T. (2011). Process and structural implications for IT-enabled outsourcing. *International Journal of Strategic Information Technology and Applications*, *2*(4), 30–43. doi:10.4018/jsita.2011100103

Dwivedi, A. N. (2009). *Handbook of research on information technology management and clinical data administration in healthcare* (Vols. 1–2). Hershey, PA: IGI Global. doi:10.4018/978-1-60566-356-2

Elbeltagi, I., McBride, N., & Hardaker, G. (2006). Evaluating the factors affecting DSS usage by senior managers in local authorities in Egypt. In M. Hunter & F. Tan (Eds.), *Advanced topics in global information management* (Vol. 5, pp. 283–307). Hershey, PA: IGI Global. doi:10.4018/978-1-59140-923-6.ch013

Eom, S., & Fountain, J. E. (2013). Enhancing information services through public-private partnerships: Information technology knowledge transfer underlying structures to develop shared services in the U.S. and Korea. In J. Gil-Garcia (Ed.), *E-government success around the world: Cases, empirical studies, and practical recommendations* (pp. 15–40). Hershey, PA: IGI Global. doi:10.4018/978-1-4666-4173-0.ch002

Esteves, T., Leuenberger, D., & Van Leuven, N. (2012). Reaching citizen 2.0: How government uses social media to send public messages during times of calm and times of crisis. In K. Kloby & M. D'Agostino (Eds.), *Citizen 2.0: Public and governmental interaction through web 2.0 technologies* (pp. 250–268). Hershey, PA: IGI Global. doi:10.4018/978-1-4666-0318-9.ch013

Estevez, E., Fillottrani, P., Janowski, T., & Ojo, A. (2012). Government information sharing: A framework for policy formulation. In Y. Chen & P. Chu (Eds.), *Electronic governance and cross-boundary collaboration: Innovations and advancing tools* (pp. 23–55). Hershey, PA: IGI Global. doi:10.4018/978-1-60960-753-1.ch002

Ezz, I. E. (2008). E-governement emerging trends: Organizational challenges. In A. Anttiroiko (Ed.), *Electronic government: Concepts, methodologies, tools, and applications* (pp. 3721–3737). Hershey, PA: IGI Global. doi:10.4018/978-1-59904-947-2.ch269

Fabri, M. (2009). The Italian style of e-justice in a comparative perspective. In A. Martínez & P. Abat (Eds.), *E-justice: Using information communication technologies in the court system* (pp. 1–19). Hershey, PA: IGI Global. doi:10.4018/978-1-59904-998-4.ch001

Fagbe, T., & Adekola, O. D. (2010). Workplace safety and personnel well-being: The impact of information technology. *International Journal of Green Computing*, *1*(1), 28–33. doi:10.4018/jgc.2010010103

Fagbe, T., & Adekola, O. D. (2011). Workplace safety and personnel well-being: The impact of information technology. In *Global business: Concepts, methodologies, tools and applications* (pp. 1438–1444). Hershey, PA: IGI Global. doi:10.4018/978-1-60960-587-2.ch509

Farmer, L. (2008). Affective collaborative instruction with librarians. In S. Kelsey & K. St.Amant (Eds.), *Handbook of research on computer mediated communication* (pp. 15–24). Hershey, PA: IGI Global. doi:10.4018/978-1-59904-863-5.ch002

Favier, L., & Mekhantar, J. (2007). Use of OSS by local e-administration: The French situation. In K. St.Amant & B. Still (Eds.), *Handbook of research on open source software: Technological, economic, and social perspectives* (pp. 428–444). Hershey, PA: IGI Global. doi:10.4018/978-1-59140-999-1.ch033

Fernando, S. (2009). Issues of e-learning in third world countries. In M. Khosrow-Pour (Ed.), *Encyclopedia of information science and technology* (2nd ed.; pp. 2273–2277). Hershey, PA: IGI Global. doi:10.4018/978-1-60566-026-4.ch360

Filho, J. R., & dos Santos, J. R. Jr. (2009). Local e-government in Brazil: Poor interaction and local politics as usual. In C. Reddick (Ed.), *Handbook of research on strategies for local e-government adoption and implementation: Comparative studies* (pp. 863–878). Hershey, PA: IGI Global. doi:10.4018/978-1-60566-282-4.ch045

Fletcher, P. D. (2004). Portals and policy: Implications of electronic access to U.S. federal government information services. In A. Pavlichev & G. Garson (Eds.), *Digital government: Principles and best practices* (pp. 52–62). Hershey, PA: IGI Global. doi:10.4018/978-1-59140-122-3.ch004

Fletcher, P. D. (2008). Portals and policy: Implications of electronic access to U.S. federal government information services. In A. Anttiroiko (Ed.), *Electronic government: Concepts, methodologies, tools, and applications* (pp. 3970–3979). Hershey, PA: IGI Global. doi:10.4018/978-1-59904-947-2.ch289

Forlano, L. (2004). The emergence of digital government: International perspectives. In A. Pavlichev & G. Garson (Eds.), *Digital government: Principles and best practices* (pp. 34–51). Hershey, PA: IGI Global. doi:10.4018/978-1-59140-122-3.ch003

Franzel, J. M., & Coursey, D. H. (2004). Government web portals: Management issues and the approaches of five states. In A. Pavlichev & G. Garson (Eds.), *Digital government: Principles and best practices* (pp. 63–77). Hershey, PA: IGI Global. doi:10.4018/978-1-59140-122-3.ch005

Gaivéo, J. M. (2013). Security of ICTs supporting healthcare activities. In M. Cruz-Cunha, I. Miranda, & P. Gonçalves (Eds.), *Handbook of research on ICTs for human-centered healthcare and social care services* (pp. 208–228). Hershey, PA: IGI Global. doi:10.4018/978-1-4666-3986-7.ch011

Garson, G. D. (1999). *Information technology and computer applications in public administration: Issues and trends*. Hershey, PA: IGI Global. doi:10.4018/978-1-87828-952-0

Garson, G. D. (2003). Toward an information technology research agenda for public administration. In G. Garson (Ed.), *Public information technology: Policy and management issues* (pp. 331–357). Hershey, PA: IGI Global. doi:10.4018/978-1-59140-060-8.ch014

Garson, G. D. (2004). The promise of digital government. In A. Pavlichev & G. Garson (Eds.), *Digital government: Principles and best practices* (pp. 2–15). Hershey, PA: IGI Global. doi:10.4018/978-1-59140-122-3.ch001

Garson, G. D. (2007). An information technology research agenda for public administration. In G. Garson (Ed.), *Modern public information technology systems: Issues and challenges* (pp. 365–392). Hershey, PA: IGI Global. doi:10.4018/978-1-59904-051-6.ch018

Gasco, M. (2007). Civil servants' resistance towards e-government development. In A. Anttiroiko & M. Malkia (Eds.), *Encyclopedia of digital government* (pp. 190–195). Hershey, PA: IGI Global. doi:10.4018/978-1-59140-789-8.ch028

Gasco, M. (2008). Civil servants' resistance towards e-government development. In A. Anttiroiko (Ed.), *Electronic government: Concepts, methodologies, tools, and applications* (pp. 2580–2588). Hershey, PA: IGI Global. doi:10.4018/978-1-59904-947-2.ch190

Ghere, R. K. (2010). Accountability and information technology enactment: Implications for social empowerment. In E. Ferro, Y. Dwivedi, J. Gil-Garcia, & M. Williams (Eds.), *Handbook of research on overcoming digital divides: Constructing an equitable and competitive information society* (pp. 515–532). Hershey, PA: IGI Global. doi:10.4018/978-1-60566-699-0.ch028

Gibson, I. W. (2012). Simulation modeling of healthcare delivery. In A. Kolker & P. Story (Eds.), *Management engineering for effective healthcare delivery: Principles and applications* (pp. 69–89). Hershey, PA: IGI Global. doi:10.4018/978-1-60960-872-9.ch003

Gil-Garcia, J. R. (2007). Exploring e-government benefits and success factors. In A. Anttiroiko & M. Malkia (Eds.), *Encyclopedia of digital government* (pp. 803–811). Hershey, PA: IGI Global. doi:10.4018/978-1-59140-789-8.ch122

Gil-Garcia, J. R., & González Miranda, F. (2010). E-government and opportunities for participation: The case of the Mexican state web portals. In C. Reddick (Ed.), *Politics, democracy and e-government: Participation and service delivery* (pp. 56–74). Hershey, PA: IGI Global. doi:10.4018/978-1-61520-933-0.ch004

Goldfinch, S. (2012). Public trust in government, trust in e-government, and use of e-government. In Z. Yan (Ed.), *Encyclopedia of cyber behavior* (pp. 987–995). Hershey, PA: IGI Global. doi:10.4018/978-1-4666-0315-8.ch081

Goodyear, M. (2012). Organizational change contributions to e-government project transitions. In S. Aikins (Ed.), *Managing e-government projects: Concepts, issues, and best practices* (pp. 1–21). Hershey, PA: IGI Global. doi:10.4018/978-1-4666-0086-7.ch001

Gordon, S., & Mulligan, P. (2003). Strategic models for the delivery of personal financial services: The role of infocracy. In S. Gordon (Ed.), *Computing information technology: The human side* (pp. 220–232). Hershey, PA: IGI Global. doi:10.4018/978-1-93177-752-0.ch014

Gordon, T. F. (2007). Legal knowledge systems. In A. Anttiroiko & M. Malkia (Eds.), *Encyclopedia of digital government* (pp. 1161–1166). Hershey, PA: IGI Global. doi:10.4018/978-1-59140-789-8.ch175

Graham, J. E., & Semich, G. W. (2008). Integrating technology to transform pedagogy: Revisiting the progress of the three phase TUI model for faculty development. In L. Tomei (Ed.), *Adapting information and communication technologies for effective education* (pp. 1–12). Hershey, PA: IGI Global. doi:10.4018/978-1-59904-922-9.ch001

Grandinetti, L., & Pisacane, O. (2012). Web services for healthcare management. In D. Prakash Vidyarthi (Ed.), *Technologies and protocols for the future of internet design: Reinventing the web* (pp. 60–94). Hershey, PA: IGI Global. doi:10.4018/978-1-4666-0203-8.ch004

Groenewegen, P., & Wagenaar, F. P. (2008). VO as an alternative to hierarchy in the Dutch police sector. In G. Putnik & M. Cruz-Cunha (Eds.), *Encyclopedia of networked and virtual organizations* (pp. 1851–1857). Hershey, PA: IGI Global. doi:10.4018/978-1-59904-885-7.ch245

Gronlund, A. (2001). Building an infrastructure to manage electronic services. In S. Dasgupta (Ed.), *Managing internet and intranet technologies in organizations: Challenges and opportunities* (pp. 71–103). Hershey, PA: IGI Global. doi:10.4018/978-1-878289-95-7.ch006

Gronlund, A. (2002). Introduction to electronic government: Design, applications and management. In Å. Grönlund (Ed.), *Electronic government: Design, applications and management* (pp. 1–21). Hershey, PA: IGI Global. doi:10.4018/978-1-930708-19-8.ch001

Gupta, A., Woosley, R., Crk, I., & Sarnikar, S. (2009). An information technology architecture for drug effectiveness reporting and post-marketing surveillance. In J. Tan (Ed.), *Medical informatics: Concepts, methodologies, tools, and applications* (pp. 631–646). Hershey, PA: IGI Global. doi:10.4018/978-1-60566-050-9.ch047

Hallin, A., & Lundevall, K. (2007). mCity: User focused development of mobile services within the city of Stockholm. In I. Kushchu (Ed.), Mobile government: An emerging direction in e-government (pp. 12-29). Hershey, PA: IGI Global. doi:10.4018/978-1-59140-884-0.ch002

Hallin, A., & Lundevall, K. (2009). mCity: User focused development of mobile services within the city of Stockholm. In S. Clarke (Ed.), Evolutionary concepts in end user productivity and performance: Applications for organizational progress (pp. 268-280). Hershey, PA: IGI Global. doi:10.4018/978-1-60566-136-0.ch017

Hallin, A., & Lundevall, K. (2009). mCity: User focused development of mobile services within the city of Stockholm. In D. Taniar (Ed.), Mobile computing: Concepts, methodologies, tools, and applications (pp. 3455-3467). Hershey, PA: IGI Global. doi:10.4018/978-1-60566-054-7.ch253

Hanson, A. (2005). Overcoming barriers in the planning of a virtual library. In M. Khosrow-Pour (Ed.), *Encyclopedia of information science and technology* (pp. 2255–2259). Hershey, PA: IGI Global. doi:10.4018/978-1-59140-553-5.ch397

Haque, A. (2008). Information technology and surveillance: Implications for public administration in a new word order. In T. Loendorf & G. Garson (Eds.), *Patriotic information systems* (pp. 177–185). Hershey, PA: IGI Global. doi:10.4018/978-1-59904-594-8.ch008

Hauck, R. V., Thatcher, S. M., & Weisband, S. P. (2012). Temporal aspects of information technology use: Increasing shift work effectiveness. In J. Wang (Ed.), *Advancing the service sector with evolving technologies: Techniques and principles* (pp. 87–104). Hershey, PA: IGI Global. doi:10.4018/978-1-4666-0044-7.ch006

Hawk, S., & Witt, T. (2006). Telecommunications courses in information systems programs. *International Journal of Information and Communication Technology Education, 2*(1), 79–92. doi:10.4018/jicte.2006010107

Helms, M. M., Moore, R., & Ahmadi, M. (2009). Information technology (IT) and the healthcare industry: A SWOT analysis. In J. Tan (Ed.), *Medical informatics: Concepts, methodologies, tools, and applications* (pp. 134–152). Hershey, PA: IGI Global. doi:10.4018/978-1-60566-050-9.ch012

Hendrickson, S. M., & Young, M. E. (2014). Electronic records management at a federally funded research and development center. In J. Krueger (Ed.), *Cases on electronic records and resource management implementation in diverse environments* (pp. 334–350). Hershey, PA: IGI Global. doi:10.4018/978-1-4666-4466-3.ch020

Henman, P. (2010). Social policy and information communication technologies. In J. Martin & L. Hawkins (Eds.), *Information communication technologies for human services education and delivery: Concepts and cases* (pp. 215–229). Hershey, PA: IGI Global. doi:10.4018/978-1-60566-735-5.ch014

Hismanoglu, M. (2011). Important issues in online education: E-pedagogy and marketing. In U. Demiray & S. Sever (Eds.), *Marketing online education programs: Frameworks for promotion and communication* (pp. 184–209). Hershey, PA: IGI Global. doi:10.4018/978-1-60960-074-7.ch012

Ho, K. K. (2008). The e-government development, IT strategies, and portals of the Hong Kong SAR government. In A. Anttiroiko (Ed.), *Electronic government: Concepts, methodologies, tools, and applications* (pp. 715–733). Hershey, PA: IGI Global. doi:10.4018/978-1-59904-947-2.ch060

Holden, S. H. (2003). The evolution of information technology management at the federal level: Implications for public administration. In G. Garson (Ed.), *Public information technology: Policy and management issues* (pp. 53–73). Hershey, PA: IGI Global. doi:10.4018/978-1-59140-060-8.ch003

Holden, S. H. (2007). The evolution of federal information technology management literature: Does IT finally matter? In G. Garson (Ed.), *Modern public information technology systems: Issues and challenges* (pp. 17–34). Hershey, PA: IGI Global. doi:10.4018/978-1-59904-051-6.ch002

Holland, J. W. (2009). Automation of American criminal justice. In M. Khosrow-Pour (Ed.), *Encyclopedia of information science and technology* (2nd ed.; pp. 300–302). Hershey, PA: IGI Global. doi:10.4018/978-1-60566-026-4.ch051

Holloway, K. (2013). Fair use, copyright, and academic integrity in an online academic environment. In *Digital rights management: Concepts, methodologies, tools, and applications* (pp. 917–928). Hershey, PA: IGI Global. doi:10.4018/978-1-4666-2136-7.ch044

Horiuchi, C. (2005). E-government databases. In L. Rivero, J. Doorn, & V. Ferraggine (Eds.), *Encyclopedia of database technologies and applications* (pp. 206–210). Hershey, PA: IGI Global. doi:10.4018/978-1-59140-560-3.ch035

Horiuchi, C. (2006). Creating IS quality in government settings. In E. Duggan & J. Reichgelt (Eds.), *Measuring information systems delivery quality* (pp. 311–327). Hershey, PA: IGI Global. doi:10.4018/978-1-59140-857-4.ch014

Hsiao, N., Chu, P., & Lee, C. (2012). Impact of e-governance on businesses: Model development and case study. In *Digital democracy: Concepts, methodologies, tools, and applications* (pp. 1407–1425). Hershey, PA: IGI Global. doi:10.4018/978-1-4666-1740-7.ch070

Huang, T., & Lee, C. (2010). Evaluating the impact of e-government on citizens: Cost-benefit analysis. In C. Reddick (Ed.), *Citizens and e-government: Evaluating policy and management* (pp. 37–52). Hershey, PA: IGI Global. doi:10.4018/978-1-61520-931-6.ch003

Hunter, M. G., Diochon, M., Pugsley, D., & Wright, B. (2002). Unique challenges for small business adoption of information technology: The case of the Nova Scotia ten. In S. Burgess (Ed.), *Managing information technology in small business: Challenges and solutions* (pp. 98–117). Hershey, PA: IGI Global. doi:10.4018/978-1-930708-35-8.ch006

Hurskainen, J. (2003). Integration of business systems and applications in merger and alliance: Case metso automation. In T. Reponen (Ed.), *Information technology enabled global customer service* (pp. 207–225). Hershey, PA: IGI Global. doi:10.4018/978-1-59140-048-6.ch012

Iazzolino, G., & Pietrantonio, R. (2011). The soveria.it project: A best practice of e-government in southern Italy. In D. Piaggesi, K. Sund, & W. Castelnovo (Eds.), *Global strategy and practice of e-governance: Examples from around the world* (pp. 34–56). Hershey, PA: IGI Global. doi:10.4018/978-1-60960-489-9.ch003

Imran, A., & Gregor, S. (2012). A process model for successful e-government adoption in the least developed countries: A case of Bangladesh. In F. Tan (Ed.), *International comparisons of information communication technologies: Advancing applications* (pp. 321–350). Hershey, PA: IGI Global. doi:10.4018/978-1-61350-480-2.ch014

Inoue, Y., & Bell, S. T. (2005). Electronic/digital government innovation, and publishing trends with IT. In M. Khosrow-Pour (Ed.), *Encyclopedia of information science and technology* (pp. 1018–1023). Hershey, PA: IGI Global. doi:10.4018/978-1-59140-553-5.ch180

Islam, M. M., & Ehsan, M. (2013). Understanding e-governance: A theoretical approach. In M. Islam & M. Ehsan (Eds.), *From government to e-governance: Public administration in the digital age* (pp. 38–49). Hershey, PA: IGI Global. doi:10.4018/978-1-4666-1909-8.ch003

Jaeger, B. (2009). E-government and e-democracy in the making. In M. Khosrow-Pour (Ed.), *Encyclopedia of information science and technology* (2nd ed.; pp. 1318–1322). Hershey, PA: IGI Global. doi:10.4018/978-1-60566-026-4.ch208

Jain, R. B. (2007). Revamping the administrative structure and processes in India for online diplomacy. In A. Anttiroiko & M. Malkia (Eds.), *Encyclopedia of digital government* (pp. 1418–1423). Hershey, PA: IGI Global. doi:10.4018/978-1-59140-789-8.ch217

Jain, R. B. (2008). Revamping the administrative structure and processes in India for online diplomacy. In A. Anttiroiko (Ed.), *Electronic government: Concepts, methodologies, tools, and applications* (pp. 3142–3149). Hershey, PA: IGI Global. doi:10.4018/978-1-59904-947-2.ch233

Jauhiainen, J. S., & Inkinen, T. (2009). E-governance and the information society in periphery. In C. Reddick (Ed.), *Handbook of research on strategies for local e-government adoption and implementation: Comparative studies* (pp. 497–514). Hershey, PA: IGI Global. doi:10.4018/978-1-60566-282-4.ch026

Jensen, M. J. (2009). Electronic democracy and citizen influence in government. In C. Reddick (Ed.), *Handbook of research on strategies for local e-government adoption and implementation: Comparative studies* (pp. 288–305). Hershey, PA: IGI Global. doi:10.4018/978-1-60566-282-4.ch015

Jiao, Y., Hurson, A. R., Potok, T. E., & Beckerman, B. G. (2009). Integrating mobile-based systems with healthcare databases. In J. Erickson (Ed.), *Database technologies: Concepts, methodologies, tools, and applications* (pp. 484–504). Hershey, PA: IGI Global. doi:10.4018/978-1-60566-058-5.ch031

Joia, L. A. (2002). A systematic model to integrate information technology into metabusinesses: A case study in the engineering realms. In F. Tan (Ed.), *Advanced topics in global information management* (Vol. 1, pp. 250–267). Hershey, PA: IGI Global. doi:10.4018/978-1-930708-43-3.ch016

Jones, T. H., & Song, I. (2000). Binary equivalents of ternary relationships in entity-relationship modeling: A logical decomposition approach. *Journal of Database Management, 11*(2), 12–19. doi:10.4018/jdm.2000040102

Juana-Espinosa, S. D. (2007). Empirical study of the municipalitites' motivations for adopting online presence. In L. Al-Hakim (Ed.), *Global e-government: Theory, applications and benchmarking* (pp. 261–279). Hershey, PA: IGI Global. doi:10.4018/978-1-59904-027-1.ch015

Jun, K., & Weare, C. (2012). Bridging from e-government practice to e-government research: Past trends and future directions. In K. Bwalya & S. Zulu (Eds.), *Handbook of research on e-government in emerging economies: Adoption, e-participation, and legal frameworks* (pp. 263–289). Hershey, PA: IGI Global. doi:10.4018/978-1-4666-0324-0.ch013

Junqueira, A., Diniz, E. H., & Fernandez, M. (2010). Electronic government implementation projects with multiple agencies: Analysis of the electronic invoice project under PMBOK framework. In J. Cordoba-Pachon & A. Ochoa-Arias (Eds.), *Systems thinking and e-participation: ICT in the governance of society* (pp. 135–153). Hershey, PA: IGI Global. doi:10.4018/978-1-60566-860-4.ch009

Juntunen, A. (2009). Joint service development with the local authorities. In C. Reddick (Ed.), *Handbook of research on strategies for local e-government adoption and implementation: Comparative studies* (pp. 902–920). Hershey, PA: IGI Global. doi:10.4018/978-1-60566-282-4.ch047

Kamel, S. (2001). *Using DSS for crisis management.* Hershey, PA: IGI Global. doi:10.4018/978-1-87828-961-2.ch020

Kamel, S. (2006). DSS for strategic decision making. In M. Khosrow-Pour (Ed.), *Cases on information technology and organizational politics & culture* (pp. 230–246). Hershey, PA: IGI Global. doi:10.4018/978-1-59904-411-8.ch015

Kamel, S. (2009). The software industry in Egypt as a potential contributor to economic growth. In M. Khosrow-Pour (Ed.), *Encyclopedia of information science and technology* (2nd ed.; pp. 3531–3537). Hershey, PA: IGI Global. doi:10.4018/978-1-60566-026-4.ch562

Kamel, S., & Hussein, M. (2008). Xceed: Pioneering the contact center industry in Egypt. *Journal of Cases on Information Technology, 10*(1), 67–91. doi:10.4018/jcit.2008010105

Kamel, S., & Wahba, K. (2003). The use of a hybrid model in web-based education: "The Global campus project. In A. Aggarwal (Ed.), *Web-based education: Learning from experience* (pp. 331–346). Hershey, PA: IGI Global. doi:10.4018/978-1-59140-102-5.ch020

Kardaras, D. K., & Papathanassiou, E. A. (2008). An exploratory study of the e-government services in Greece. In G. Garson & M. Khosrow-Pour (Eds.), *Handbook of research on public information technology* (pp. 162–174). Hershey, PA: IGI Global. doi:10.4018/978-1-59904-857-4.ch016

Kassahun, A. E., Molla, A., & Sarkar, P. (2012). Government process reengineering: What we know and what we need to know. In *Digital democracy: Concepts, methodologies, tools, and applications* (pp. 1730–1752). Hershey, PA: IGI Global. doi:10.4018/978-1-4666-1740-7.ch086

Khan, B. (2005). Technological issues. In B. Khan (Ed.), *Managing e-learning strategies: Design, delivery, implementation and evaluation* (pp. 154–180). Hershey, PA: IGI Global. doi:10.4018/978-1-59140-634-1.ch004

Khasawneh, A., Bsoul, M., Obeidat, I., & Al Azzam, I. (2012). Technology fears: A study of e-commerce loyalty perception by Jordanian customers. In J. Wang (Ed.), *Advancing the service sector with evolving technologies: Techniques and principles* (pp. 158–165). Hershey, PA: IGI Global. doi:10.4018/978-1-4666-0044-7.ch010

Khatibi, V., & Montazer, G. A. (2012). E-research methodology. In A. Juan, T. Daradoumis, M. Roca, S. Grasman, & J. Faulin (Eds.), *Collaborative and distributed e-research: Innovations in technologies, strategies and applications* (pp. 62–81). Hershey, PA: IGI Global. doi:10.4018/978-1-4666-0125-3.ch003

Kidd, T. (2011). The dragon in the school's backyard: A review of literature on the uses of technology in urban schools. In L. Tomei (Ed.), *Online courses and ICT in education: Emerging practices and applications* (pp. 242–257). Hershey, PA: IGI Global. doi:10.4018/978-1-60960-150-8.ch019

Kidd, T. T. (2010). My experience tells the story: Exploring technology adoption from a qualitative perspective - A pilot study. In H. Song & T. Kidd (Eds.), *Handbook of research on human performance and instructional technology* (pp. 247–262). Hershey, PA: IGI Global. doi:10.4018/978-1-60566-782-9.ch015

Kieley, B., Lane, G., Paquet, G., & Roy, J. (2002). e-Government in Canada: Services online or public service renewal? In Å. Grönlund (Ed.), Electronic government: Design, applications and management (pp. 340-355). Hershey, PA: IGI Global. doi:10.4018/978-1-930708-19-8.ch016

Kim, P. (2012). "Stay out of the way! My kid is video blogging through a phone!": A lesson learned from math tutoring social media for children in underserved communities. In *Wireless technologies: Concepts, methodologies, tools and applications* (pp. 1415–1428). Hershey, PA: IGI Global. doi:10.4018/978-1-61350-101-6.ch517

Kirlidog, M. (2010). Financial aspects of national ICT strategies. In S. Kamel (Ed.), *E-strategies for technological diffusion and adoption: National ICT approaches for socioeconomic development* (pp. 277–292). Hershey, PA: IGI Global. doi:10.4018/978-1-60566-388-3.ch016

Kisielnicki, J. (2006). Transfer of information and knowledge in the project management. In E. Coakes & S. Clarke (Eds.), *Encyclopedia of communities of practice in information and knowledge management* (pp. 544–551). Hershey, PA: IGI Global. doi:10.4018/978-1-59140-556-6.ch091

Kittner, M., & Van Slyke, C. (2006). Reorganizing information technology services in an academic environment. In M. Khosrow-Pour (Ed.), *Cases on the human side of information technology* (pp. 49–66). Hershey, PA: IGI Global. doi:10.4018/978-1-59904-405-7.ch004

Knoell, H. D. (2008). Semi virtual workplaces in German financial service enterprises. In P. Zemliansky & K. St.Amant (Eds.), *Handbook of research on virtual workplaces and the new nature of business practices* (pp. 570–581). Hershey, PA: IGI Global. doi:10.4018/978-1-59904-893-2.ch041

Koh, S. L., & Maguire, S. (2009). Competing in the age of information technology in a developing economy: Experiences of an Indian bank. In S. Koh & S. Maguire (Eds.), *Information and communication technologies management in turbulent business environments* (pp. 326–350). Hershey, PA: IGI Global. doi:10.4018/978-1-60566-424-8.ch018

Kollmann, T., & Häsel, M. (2009). Competence of information technology professionals in internet-based ventures. In I. Lee (Ed.), *Electronic business: Concepts, methodologies, tools, and applications* (pp. 1905–1919). Hershey, PA: IGI Global. doi:10.4018/978-1-60566-056-1.ch118

Kollmann, T., & Häsel, M. (2009). Competence of information technology professionals in internet-based ventures. In A. Cater-Steel (Ed.), *Information technology governance and service management: Frameworks and adaptations* (pp. 239–253). Hershey, PA: IGI Global. doi:10.4018/978-1-60566-008-0.ch013

Kollmann, T., & Häsel, M. (2010). Competence of information technology professionals in internet-based ventures. In *Electronic services: Concepts, methodologies, tools and applications* (pp. 1551–1565). Hershey, PA: IGI Global. doi:10.4018/978-1-61520-967-5.ch094

Kraemer, K., & King, J. L. (2006). Information technology and administrative reform: Will e-government be different? *International Journal of Electronic Government Research*, *2*(1), 1–20. doi:10.4018/jegr.2006010101

Kraemer, K., & King, J. L. (2008). Information technology and administrative reform: Will e-government be different? In D. Norris (Ed.), *E-government research: Policy and management* (pp. 1–20). Hershey, PA: IGI Global. doi:10.4018/978-1-59904-913-7.ch001

Lampathaki, F., Tsiakaliaris, C., Stasis, A., & Charalabidis, Y. (2011). National interoperability frameworks: The way forward. In Y. Charalabidis (Ed.), *Interoperability in digital public services and administration: Bridging e-government and e-business* (pp. 1–24). Hershey, PA: IGI Global. doi:10.4018/978-1-61520-887-6.ch001

Lan, Z., & Scott, C. R. (1996). The relative importance of computer-mediated information versus conventional non-computer-mediated information in public managerial decision making. *Information Resources Management Journal, 9*(1), 27–0. doi:10.4018/irmj.1996010103

Law, W. (2004). *Public sector data management in a developing economy*. Hershey, PA: IGI Global. doi:10.4018/978-1-59140-259-6.ch034

Law, W. K. (2005). Information resources development challenges in a cross-cultural environment. In M. Khosrow-Pour (Ed.), *Encyclopedia of information science and technology* (pp. 1476–1481). Hershey, PA: IGI Global. doi:10.4018/978-1-59140-553-5.ch259

Law, W. K. (2009). Cross-cultural challenges for information resources management. In M. Khosrow-Pour (Ed.), *Encyclopedia of information science and technology* (2nd ed.; pp. 840–846). Hershey, PA: IGI Global. doi:10.4018/978-1-60566-026-4.ch136

Law, W. K. (2011). Cross-cultural challenges for information resources management. In *Global business: Concepts, methodologies, tools and applications* (pp. 1924–1932). Hershey, PA: IGI Global. doi:10.4018/978-1-60960-587-2.ch704

Malkia, M., & Savolainen, R. (2004). eTransformation in government, politics and society: Conceptual framework and introduction. In M. Malkia, A. Anttiroiko, & R. Savolainen (Eds.), eTransformation in governance: New directions in government and politics (pp. 1-21). Hershey, PA: IGI Global. doi:10.4018/978-1-59140-130-8.ch001

Mandujano, S. (2011). Network manageability security. In D. Kar & M. Syed (Eds.), *Network security, administration and management: Advancing technology and practice* (pp. 158–181). Hershey, PA: IGI Global. doi:10.4018/978-1-60960-777-7.ch009

Marich, M. J., Schooley, B. L., & Horan, T. A. (2012). A normative enterprise architecture for guiding end-to-end emergency response decision support. In M. Jennex (Ed.), *Managing crises and disasters with emerging technologies: Advancements* (pp. 71–87). Hershey, PA: IGI Global. doi:10.4018/978-1-4666-0167-3.ch006

Markov, R., & Okujava, S. (2008). Costs, benefits, and risks of e-government portals. In G. Putnik & M. Cruz-Cunha (Eds.), *Encyclopedia of networked and virtual organizations* (pp. 354–363). Hershey, PA: IGI Global. doi:10.4018/978-1-59904-885-7.ch047

Martin, N., & Rice, J. (2013). Evaluating and designing electronic government for the future: Observations and insights from Australia. In V. Weerakkody (Ed.), *E-government services design, adoption, and evaluation* (pp. 238–258). Hershey, PA: IGI Global. doi:10.4018/978-1-4666-2458-0.ch014

i. Martinez, A. C. (2008). Accessing administration's information via internet in Spain. In F. Tan (Ed.), *Global information technologies: Concepts, methodologies, tools, and applications* (pp. 2558–2573). Hershey, PA: IGI Global. doi:10.4018/978-1-59904-939-7.ch186

Mbarika, V. W., Meso, P. N., & Musa, P. F. (2006). A disconnect in stakeholders' perceptions from emerging realities of teledensity growth in Africa's least developed countries. In M. Hunter & F. Tan (Eds.), *Advanced topics in global information management* (Vol. 5, pp. 263–282). Hershey, PA: IGI Global. doi:10.4018/978-1-59140-923-6.ch012

Mbarika, V. W., Meso, P. N., & Musa, P. F. (2008). A disconnect in stakeholders' perceptions from emerging realities of teledensity growth in Africa's least developed countries. In F. Tan (Ed.), *Global information technologies: Concepts, methodologies, tools, and applications* (pp. 2948–2962). Hershey, PA: IGI Global. doi:10.4018/978-1-59904-939-7.ch209

Means, T., Olson, E., & Spooner, J. (2013). Discovering ways that don't work on the road to success: Strengths and weaknesses revealed by an active learning studio classroom project. In A. Benson, J. Moore, & S. Williams van Rooij (Eds.), *Cases on educational technology planning, design, and implementation: A project management perspective* (pp. 94–113). Hershey, PA: IGI Global. doi:10.4018/978-1-4666-4237-9.ch006

Melitski, J., Holzer, M., Kim, S., Kim, C., & Rho, S. (2008). Digital government worldwide: An e-government assessment of municipal web sites. In G. Garson & M. Khosrow-Pour (Eds.), *Handbook of research on public information technology* (pp. 790–804). Hershey, PA: IGI Global. doi:10.4018/978-1-59904-857-4.ch069

Memmola, M., Palumbo, G., & Rossini, M. (2009). Web & RFID technology: New frontiers in costing and process management for rehabilitation medicine. In L. Al-Hakim & M. Memmola (Eds.), *Business web strategy: Design, alignment, and application* (pp. 145–169). Hershey, PA: IGI Global. doi:10.4018/978-1-60566-024-0.ch008

Meng, Z., Fahong, Z., & Lei, L. (2008). Information technology and environment. In Y. Kurihara, S. Takaya, H. Harui, & H. Kamae (Eds.), *Information technology and economic development* (pp. 201–212). Hershey, PA: IGI Global. doi:10.4018/978-1-59904-579-5.ch014

Mentzingen de Moraes, A. J., Ferneda, E., Costa, I., & Spinola, M. D. (2011). Practical approach for implementation of governance process in IT: Information technology areas. In N. Shi & G. Silvius (Eds.), *Enterprise IT governance, business value and performance measurement* (pp. 19–40). Hershey, PA: IGI Global. doi:10.4018/978-1-60566-346-3.ch002

Merwin, G. A. Jr, McDonald, J. S., & Odera, L. C. (2008). Economic development: Government's cutting edge in IT. In M. Raisinghani (Ed.), *Handbook of research on global information technology management in the digital economy* (pp. 1–37). Hershey, PA: IGI Global. doi:10.4018/978-1-59904-875-8.ch001

Meso, P., & Duncan, N. (2002). Can national information infrastructures enhance social development in the least developed countries? An empirical investigation. In M. Dadashzadeh (Ed.), *Information technology management in developing countries* (pp. 23–51). Hershey, PA: IGI Global. doi:10.4018/978-1-931777-03-2.ch002

Meso, P. N., & Duncan, N. B. (2002). Can national information infrastructures enhance social development in the least developed countries? In F. Tan (Ed.), *Advanced topics in global information management* (Vol. 1, pp. 207–226). Hershey, PA: IGI Global. doi:10.4018/978-1-930708-43-3.ch014

Middleton, M. (2008). Evaluation of e-government web sites. In G. Garson & M. Khosrow-Pour (Eds.), *Handbook of research on public information technology* (pp. 699–710). Hershey, PA: IGI Global. doi:10.4018/978-1-59904-857-4.ch063

Mingers, J. (2010). Pluralism, realism, and truth: The keys to knowledge in information systems research. In D. Paradice (Ed.), *Emerging systems approaches in information technologies: Concepts, theories, and applications* (pp. 86–98). Hershey, PA: IGI Global. doi:10.4018/978-1-60566-976-2.ch006

Mital, K. M. (2012). ICT, unique identity and inclusive growth: An Indian perspective. In A. Manoharan & M. Holzer (Eds.), *E-governance and civic engagement: Factors and determinants of e-democracy* (pp. 584–612). Hershey, PA: IGI Global. doi:10.4018/978-1-61350-083-5.ch029

Mizell, A. P. (2008). Helping close the digital divide for financially disadvantaged seniors. In F. Tan (Ed.), *Global information technologies: Concepts, methodologies, tools, and applications* (pp. 2396–2402). Hershey, PA: IGI Global. doi:10.4018/978-1-59904-939-7.ch173

Molinari, F., Wills, C., Koumpis, A., & Moumtzi, V. (2011). A citizen-centric platform to support networking in the area of e-democracy. In H. Rahman (Ed.), *Cases on adoption, diffusion and evaluation of global e-governance systems: Impact at the grass roots* (pp. 282–302). Hershey, PA: IGI Global. doi:10.4018/978-1-61692-814-8.ch014

Molinari, F., Wills, C., Koumpis, A., & Moumtzi, V. (2013). A citizen-centric platform to support networking in the area of e-democracy. In H. Rahman (Ed.), *Cases on progressions and challenges in ICT utilization for citizen-centric governance* (pp. 265–297). Hershey, PA: IGI Global. doi:10.4018/978-1-4666-2071-1.ch013

Monteverde, F. (2010). The process of e-government public policy inclusion in the governmental agenda: A framework for assessment and case study. In J. Cordoba-Pachon & A. Ochoa-Arias (Eds.), *Systems thinking and e-participation: ICT in the governance of society* (pp. 233–245). Hershey, PA: IGI Global. doi:10.4018/978-1-60566-860-4.ch015

Moodley, S. (2008). Deconstructing the South African government's ICT for development discourse. In A. Anttiroiko (Ed.), *Electronic government: Concepts, methodologies, tools, and applications* (pp. 622–631). Hershey, PA: IGI Global. doi:10.4018/978-1-59904-947-2.ch053

Moodley, S. (2008). Deconstructing the South African government's ICT for development discourse. In C. Van Slyke (Ed.), *Information communication technologies: Concepts, methodologies, tools, and applications* (pp. 816–825). Hershey, PA: IGI Global. doi:10.4018/978-1-59904-949-6.ch052

Mora, M., Cervantes-Perez, F., Gelman-Muravchik, O., Forgionne, G. A., & Mejia-Olvera, M. (2003). DMSS implementation research: A conceptual analysis of the contributions and limitations of the factor-based and stage-based streams. In G. Forgionne, J. Gupta, & M. Mora (Eds.), *Decision-making support systems: Achievements and challenges for the new decade* (pp. 331–356). Hershey, PA: IGI Global. doi:10.4018/978-1-59140-045-5.ch020

Mörtberg, C., & Elovaara, P. (2010). Attaching people and technology: Between e and government. In S. Booth, S. Goodman, & G. Kirkup (Eds.), *Gender issues in learning and working with information technology: Social constructs and cultural contexts* (pp. 83–98). Hershey, PA: IGI Global. doi:10.4018/978-1-61520-813-5.ch005

Murphy, J., Harper, E., Devine, E. C., Burke, L. J., & Hook, M. L. (2011). Case study: Lessons learned when embedding evidence-based knowledge in a nurse care planning and documentation system. In A. Cashin & R. Cook (Eds.), *Evidence-based practice in nursing informatics: Concepts and applications* (pp. 174–190). Hershey, PA: IGI Global. doi:10.4018/978-1-60960-034-1.ch014

Mutula, S. M. (2013). E-government's role in poverty alleviation: Case study of South Africa. In H. Rahman (Ed.), *Cases on progressions and challenges in ICT utilization for citizen-centric governance* (pp. 44–68). Hershey, PA: IGI Global. doi:10.4018/978-1-4666-2071-1.ch003

Nath, R., & Angeles, R. (2005). Relationships between supply characteristics and buyer-supplier coupling in e-procurement: An empirical analysis. *International Journal of E-Business Research, 1*(2), 40–55. doi:10.4018/jebr.2005040103

Nissen, M. E. (2006). Application cases in government. In M. Nissen (Ed.), *Harnessing knowledge dynamics: Principled organizational knowing & learning* (pp. 152–181). Hershey, PA: IGI Global. doi:10.4018/978-1-59140-773-7.ch008

Norris, D. F. (2003). Leading-edge information technologies and American local governments. In G. Garson (Ed.), *Public information technology: Policy and management issues* (pp. 139–169). Hershey, PA: IGI Global. doi:10.4018/978-1-59140-060-8.ch007

Norris, D. F. (2008). Information technology among U.S. local governments. In G. Garson & M. Khosrow-Pour (Eds.), *Handbook of research on public information technology* (pp. 132–144). Hershey, PA: IGI Global. doi:10.4018/978-1-59904-857-4.ch013

Northrop, A. (1999). The challenge of teaching information technology in public administration graduate programs. In G. Garson (Ed.), *Information technology and computer applications in public administration: Issues and trends* (pp. 1–22). Hershey, PA: IGI Global. doi:10.4018/978-1-87828-952-0.ch001

Northrop, A. (2003). Information technology and public administration: The view from the profession. In G. Garson (Ed.), *Public information technology: Policy and management issues* (pp. 1–19). Hershey, PA: IGI Global. doi:10.4018/978-1-59140-060-8.ch001

Northrop, A. (2007). Lip service? How PA journals and textbooks view information technology. In G. Garson (Ed.), *Modern public information technology systems: Issues and challenges* (pp. 1–16). Hershey, PA: IGI Global. doi:10.4018/978-1-59904-051-6.ch001

Null, E. (2013). Legal and political barriers to municipal networks in the United States. In A. Abdelaal (Ed.), *Social and economic effects of community wireless networks and infrastructures* (pp. 27–56). Hershey, PA: IGI Global. doi:10.4018/978-1-4666-2997-4.ch003

Okunoye, A., Frolick, M., & Crable, E. (2006). ERP implementation in higher education: An account of pre-implementation and implementation phases. *Journal of Cases on Information Technology*, *8*(2), 110–132. doi:10.4018/jcit.2006040106

Olasina, G. (2012). A review of egovernment services in Nigeria. In A. Tella & A. Issa (Eds.), *Library and information science in developing countries: Contemporary issues* (pp. 205–221). Hershey, PA: IGI Global. doi:10.4018/978-1-61350-335-5.ch015

Orgeron, C. P. (2008). A model for reengineering IT job classes in state government. In G. Garson & M. Khosrow-Pour (Eds.), *Handbook of research on public information technology* (pp. 735–746). Hershey, PA: IGI Global. doi:10.4018/978-1-59904-857-4.ch066

Owsinski, J. W., & Pielak, A. M. (2011). Local authority websites in rural areas: Measuring quality and functionality, and assessing the role. In Z. Andreopoulou, B. Manos, N. Polman, & D. Viaggi (Eds.), *Agricultural and environmental informatics, governance and management: Emerging research applications* (pp. 39–60). Hershey, PA: IGI Global. doi:10.4018/978-1-60960-621-3.ch003

Owsiński, J. W., Pielak, A. M., Sęp, K., & Stańczak, J. (2014). Local web-based networks in rural municipalities: Extension, density, and meaning. In Z. Andreopoulou, V. Samathrakis, S. Louca, & M. Vlachopoulou (Eds.), *E-innovation for sustainable development of rural resources during global economic crisis* (pp. 126–151). Hershey, PA: IGI Global. doi:10.4018/978-1-4666-4550-9.ch011

Pagani, M., & Pasinetti, C. (2008). Technical and functional quality in the development of t-government services. In A. Anttiroiko (Ed.), *Electronic government: Concepts, methodologies, tools, and applications* (pp. 2943–2965). Hershey, PA: IGI Global. doi:10.4018/978-1-59904-947-2.ch220

Pani, A. K., & Agrahari, A. (2005). On e-markets in emerging economy: An Indian experience. In M. Khosrow-Pour (Ed.), *Advanced topics in electronic commerce* (Vol. 1, pp. 287–299). Hershey, PA: IGI Global. doi:10.4018/978-1-59140-819-2.ch015

Papadopoulos, T., Angelopoulos, S., & Kitsios, F. (2011). A strategic approach to e-health interoperability using e-government frameworks. In A. Lazakidou, K. Siassiakos, & K. Ioannou (Eds.), *Wireless technologies for ambient assisted living and healthcare: Systems and applications* (pp. 213–229). Hershey, PA: IGI Global. doi:10.4018/978-1-61520-805-0.ch012

Papadopoulos, T., Angelopoulos, S., & Kitsios, F. (2013). A strategic approach to e-health interoperability using e-government frameworks. In *User-driven healthcare: Concepts, methodologies, tools, and applications* (pp. 791–807). Hershey, PA: IGI Global. doi:10.4018/978-1-4666-2770-3.ch039

Papaleo, G., Chiarella, D., Aiello, M., & Caviglione, L. (2012). Analysis, development and deployment of statistical anomaly detection techniques for real e-mail traffic. In T. Chou (Ed.), *Information assurance and security technologies for risk assessment and threat management: Advances* (pp. 47–71). Hershey, PA: IGI Global. doi:10.4018/978-1-61350-507-6.ch003

Papp, R. (2003). Information technology & FDA compliance in the pharmaceutical industry. In M. Khosrow-Pour (Ed.), *Annals of cases on information technology* (Vol. 5, pp. 262–273). Hershey, PA: IGI Global. doi:10.4018/978-1-59140-061-5.ch017

Parsons, T. W. (2007). Developing a knowledge management portal. In A. Tatnall (Ed.), *Encyclopedia of portal technologies and applications* (pp. 223–227). Hershey, PA: IGI Global. doi:10.4018/978-1-59140-989-2.ch039

Passaris, C. E. (2007). Immigration and digital government. In A. Anttiroiko & M. Malkia (Eds.), *Encyclopedia of digital government* (pp. 988–994). Hershey, PA: IGI Global. doi:10.4018/978-1-59140-789-8.ch148

Pavlichev, A. (2004). The e-government challenge for public administration. In A. Pavlichev & G. Garson (Eds.), *Digital government: Principles and best practices* (pp. 276–290). Hershey, PA: IGI Global. doi:10.4018/978-1-59140-122-3.ch018

Penrod, J. I., & Harbor, A. F. (2000). Designing and implementing a learning organization-oriented information technology planning and management process. In L. Petrides (Ed.), *Case studies on information technology in higher education: Implications for policy and practice* (pp. 7–19). Hershey, PA: IGI Global. doi:10.4018/978-1-878289-74-2.ch001

Planas-Silva, M. D., & Joseph, R. C. (2011). Perspectives on the adoption of electronic resources for use in clinical trials. In M. Guah (Ed.), *Healthcare delivery reform and new technologies: Organizational initiatives* (pp. 19–28). Hershey, PA: IGI Global. doi:10.4018/978-1-60960-183-6.ch002

Pomazalová, N., & Rejman, S. (2013). The rationale behind implementation of new electronic tools for electronic public procurement. In N. Pomazalová (Ed.), *Public sector transformation processes and internet public procurement: Decision support systems* (pp. 85–117). Hershey, PA: IGI Global. doi:10.4018/978-1-4666-2665-2.ch006

Postorino, M. N. (2012). City competitiveness and airport: Information science perspective. In M. Bulu (Ed.), *City competitiveness and improving urban subsystems: Technologies and applications* (pp. 61–83). Hershey, PA: IGI Global. doi:10.4018/978-1-61350-174-0.ch004

Poupa, C. (2002). Electronic government in Switzerland: Priorities for 2001-2005 - Electronic voting and federal portal. In Å. Grönlund (Ed.), *Electronic government: Design, applications and management* (pp. 356–369). Hershey, PA: IGI Global. doi:10.4018/978-1-930708-19-8.ch017

Powell, S. R. (2010). Interdisciplinarity in telecommunications and networking. In *Networking and telecommunications: Concepts, methodologies, tools and applications* (pp. 33–40). Hershey, PA: IGI Global. doi:10.4018/978-1-60566-986-1.ch004

Priya, P. S., & Mathiyalagan, N. (2011). A study of the implementation status of two e-governance projects in land revenue administration in India. In M. Shareef, V. Kumar, U. Kumar, & Y. Dwivedi (Eds.), *Stakeholder adoption of e-government services: Driving and resisting factors* (pp. 214–230). Hershey, PA: IGI Global. doi:10.4018/978-1-60960-601-5.ch011

Prysby, C., & Prysby, N. (2000). Electronic mail, employee privacy and the workplace. In L. Janczewski (Ed.), *Internet and intranet security management: Risks and solutions* (pp. 251–270). Hershey, PA: IGI Global. doi:10.4018/978-1-878289-71-1.ch009

Prysby, C. L., & Prysby, N. D. (2003). Electronic mail in the public workplace: Issues of privacy and public disclosure. In G. Garson (Ed.), *Public information technology: Policy and management issues* (pp. 271–298). Hershey, PA: IGI Global. doi:10.4018/978-1-59140-060-8.ch012

Prysby, C. L., & Prysby, N. D. (2007). You have mail, but who is reading it? Issues of e-mail in the public workplace. In G. Garson (Ed.), *Modern public information technology systems: Issues and challenges* (pp. 312–336). Hershey, PA: IGI Global. doi:10.4018/978-1-59904-051-6.ch016

Radl, A., & Chen, Y. (2005). Computer security in electronic government: A state-local education information system. *International Journal of Electronic Government Research, 1*(1), 79–99. doi:10.4018/jegr.2005010105

Rahman, H. (2008). Information dynamics in developing countries. In C. Van Slyke (Ed.), *Information communication technologies: Concepts, methodologies, tools, and applications* (pp. 104–114). Hershey, PA: IGI Global. doi:10.4018/978-1-59904-949-6.ch008

Ramanathan, J. (2009). Adaptive IT architecture as a catalyst for network capability in government. In P. Saha (Ed.), *Advances in government enterprise architecture* (pp. 149–172). Hershey, PA: IGI Global. doi:10.4018/978-1-60566-068-4.ch007

Ramos, I., & Berry, D. M. (2006). Social construction of information technology supporting work. In M. Khosrow-Pour (Ed.), *Cases on information technology: Lessons learned* (Vol. 7, pp. 36–52). Hershey, PA: IGI Global. doi:10.4018/978-1-59140-673-0.ch003

Ray, D., Gulla, U., Gupta, M. P., & Dash, S. S. (2009). Interoperability and constituents of interoperable systems in public sector. In V. Weerakkody, M. Janssen, & Y. Dwivedi (Eds.), *Handbook of research on ICT-enabled transformational government: A global perspective* (pp. 175–195). Hershey, PA: IGI Global. doi:10.4018/978-1-60566-390-6.ch010

Reddick, C. G. (2007). E-government and creating a citizen-centric government: A study of federal government CIOs. In G. Garson (Ed.), *Modern public information technology systems: Issues and challenges* (pp. 143–165). Hershey, PA: IGI Global. doi:10.4018/978-1-59904-051-6.ch008

Reddick, C. G. (2010). Citizen-centric e-government. In C. Reddick (Ed.), *Homeland security preparedness and information systems: Strategies for managing public policy* (pp. 45–75). Hershey, PA: IGI Global. doi:10.4018/978-1-60566-834-5.ch002

Reddick, C. G. (2010). E-government and creating a citizen-centric government: A study of federal government CIOs. In C. Reddick (Ed.), *Homeland security preparedness and information systems: Strategies for managing public policy* (pp. 230–250). Hershey, PA: IGI Global. doi:10.4018/978-1-60566-834-5.ch012

Reddick, C. G. (2010). Perceived effectiveness of e-government and its usage in city governments: Survey evidence from information technology directors. In C. Reddick (Ed.), *Homeland security preparedness and information systems: Strategies for managing public policy* (pp. 213–229). Hershey, PA: IGI Global. doi:10.4018/978-1-60566-834-5.ch011

Reddick, C. G. (2012). Customer relationship management adoption in local governments in the United States. In S. Chhabra & M. Kumar (Eds.), *Strategic enterprise resource planning models for e-government: Applications and methodologies* (pp. 111–124). Hershey, PA: IGI Global. doi:10.4018/978-1-60960-863-7.ch008

Reeder, F. S., & Pandy, S. M. (2008). Identifying effective funding models for e-government. In A. Anttiroiko (Ed.), *Electronic government: Concepts, methodologies, tools, and applications* (pp. 1108–1138). Hershey, PA: IGI Global. doi:10.4018/978-1-59904-947-2.ch083

Riesco, D., Acosta, E., & Montejano, G. (2003). An extension to a UML activity graph from workflow. In L. Favre (Ed.), *UML and the unified process* (pp. 294–314). Hershey, PA: IGI Global. doi:10.4018/978-1-93177-744-5.ch015

Ritzhaupt, A. D., & Gill, T. G. (2008). A hybrid and novel approach to teaching computer programming in MIS curriculum. In S. Negash, M. Whitman, A. Woszczynski, K. Hoganson, & H. Mattord (Eds.), *Handbook of distance learning for real-time and asynchronous information technology education* (pp. 259–281). Hershey, PA: IGI Global. doi:10.4018/978-1-59904-964-9.ch014

Roche, E. M. (1993). International computing and the international regime. *Journal of Global Information Management, 1*(2), 33–44. doi:10.4018/jgim.1993040103

Rocheleau, B. (2007). Politics, accountability, and information management. In G. Garson (Ed.), *Modern public information technology systems: Issues and challenges* (pp. 35–71). Hershey, PA: IGI Global. doi:10.4018/978-1-59904-051-6.ch003

Rodrigues Filho, J. (2010). E-government in Brazil: Reinforcing dominant institutions or reducing citizenship? In C. Reddick (Ed.), *Politics, democracy and e-government: Participation and service delivery* (pp. 347–362). Hershey, PA: IGI Global. doi:10.4018/978-1-61520-933-0.ch021

Rodriguez, S. R., & Thorp, D. A. (2013). eLearning for industry: A case study of the project management process. In A. Benson, J. Moore, & S. Williams van Rooij (Eds.), Cases on educational technology planning, design, and implementation: A project management perspective (pp. 319-342). Hershey, PA: IGI Global. doi:10.4018/978-1-4666-4237-9.ch017

Roman, A. V. (2013). Delineating three dimensions of e-government success: Security, functionality, and transformation. In J. Gil-Garcia (Ed.), *E-government success factors and measures: Theories, concepts, and methodologies* (pp. 171–192). Hershey, PA: IGI Global. doi:10.4018/978-1-4666-4058-0.ch010

Ross, S. C., Tyran, C. K., & Auer, D. J. (2008). Up in smoke: Rebuilding after an IT disaster. In H. Nemati (Ed.), *Information security and ethics: Concepts, methodologies, tools, and applications* (pp. 3659–3675). Hershey, PA: IGI Global. doi:10.4018/978-1-59904-937-3.ch248

Ross, S. C., Tyran, C. K., Auer, D. J., Junell, J. M., & Williams, T. G. (2005). Up in smoke: Rebuilding after an IT disaster. *Journal of Cases on Information Technology*, *7*(2), 31–49. doi:10.4018/jcit.2005040103

Roy, J. (2008). Security, sovereignty, and continental interoperability: Canada's elusive balance. In T. Loendorf & G. Garson (Eds.), *Patriotic information systems* (pp. 153–176). Hershey, PA: IGI Global. doi:10.4018/978-1-59904-594-8.ch007

Rubeck, R. F., & Miller, G. A. (2009). vGOV: Remote video access to government services. In A. Scupola (Ed.), Cases on managing e-services (pp. 253-268). Hershey, PA: IGI Global. doi:10.4018/978-1-60566-064-6.ch017

Saekow, A., & Boonmee, C. (2011). The challenges of implementing e-government interoperability in Thailand: Case of official electronic correspondence letters exchange across government departments. In Y. Charalabidis (Ed.), *Interoperability in digital public services and administration: Bridging e-government and e-business* (pp. 40–61). Hershey, PA: IGI Global. doi:10.4018/978-1-61520-887-6.ch003

Saekow, A., & Boonmee, C. (2012). The challenges of implementing e-government interoperability in Thailand: Case of official electronic correspondence letters exchange across government departments. In *Digital democracy: Concepts, methodologies, tools, and applications* (pp. 1883–1905). Hershey, PA: IGI Global. doi:10.4018/978-1-4666-1740-7.ch094

Sagsan, M., & Medeni, T. (2012). Understanding "knowledge management (KM) paradigms" from social media perspective: An empirical study on discussion group for KM at professional networking site. In M. Cruz-Cunha, P. Gonçalves, N. Lopes, E. Miranda, & G. Putnik (Eds.), *Handbook of research on business social networking: Organizational, managerial, and technological dimensions* (pp. 738–755). Hershey, PA: IGI Global. doi:10.4018/978-1-61350-168-9.ch039

Sahi, G., & Madan, S. (2013). Information security threats in ERP enabled e-governance: Challenges and solutions. In *Enterprise resource planning: Concepts, methodologies, tools, and applications* (pp. 825–837). Hershey, PA: IGI Global. doi:10.4018/978-1-4666-4153-2.ch048

Sanford, C., & Bhattacherjee, A. (2008). IT implementation in a developing country municipality: A sociocognitive analysis. *International Journal of Technology and Human Interaction, 4*(3), 68–93. doi:10.4018/jthi.2008070104

Schelin, S. H. (2003). E-government: An overview. In G. Garson (Ed.), *Public information technology: Policy and management issues* (pp. 120–138). Hershey, PA: IGI Global. doi:10.4018/978-1-59140-060-8.ch006

Schelin, S. H. (2004). Training for digital government. In A. Pavlichev & G. Garson (Eds.), *Digital government: Principles and best practices* (pp. 263–275). Hershey, PA: IGI Global. doi:10.4018/978-1-59140-122-3.ch017

Schelin, S. H. (2007). E-government: An overview. In G. Garson (Ed.), *Modern public information technology systems: Issues and challenges* (pp. 110–126). Hershey, PA: IGI Global. doi:10.4018/978-1-59904-051-6.ch006

Schelin, S. H., & Garson, G. (2004). Theoretical justification of critical success factors. In G. Garson & S. Schelin (Eds.), *IT solutions series: Humanizing information technology: Advice from experts* (pp. 4–15). Hershey, PA: IGI Global. doi:10.4018/978-1-59140-245-9.ch002

Scime, A. (2002). Information systems and computer science model curricula: A comparative look. In M. Dadashzadeh, A. Saber, & S. Saber (Eds.), *Information technology education in the new millennium* (pp. 146–158). Hershey, PA: IGI Global. doi:10.4018/978-1-931777-05-6.ch018

Scime, A. (2009). Computing curriculum analysis and development. In M. Khosrow-Pour (Ed.), *Encyclopedia of information science and technology* (2nd ed.; pp. 667–671). Hershey, PA: IGI Global. doi:10.4018/978-1-60566-026-4.ch108

Scime, A., & Wania, C. (2008). Computing curricula: A comparison of models. In C. Van Slyke (Ed.), *Information communication technologies: Concepts, methodologies, tools, and applications* (pp. 1270–1283). Hershey, PA: IGI Global. doi:10.4018/978-1-59904-949-6.ch088

Seidman, S. B. (2009). An international perspective on professional software engineering credentials. In H. Ellis, S. Demurjian, & J. Naveda (Eds.), *Software engineering: Effective teaching and learning approaches and practices* (pp. 351–361). Hershey, PA: IGI Global. doi:10.4018/978-1-60566-102-5.ch018

Seifert, J. W. (2007). E-government act of 2002 in the United States. In A. Anttiroiko & M. Malkia (Eds.), *Encyclopedia of digital government* (pp. 476–481). Hershey, PA: IGI Global. doi:10.4018/978-1-59140-789-8.ch072

Seifert, J. W., & Relyea, H. C. (2008). E-government act of 2002 in the United States. In A. Anttiroiko (Ed.), *Electronic government: Concepts, methodologies, tools, and applications* (pp. 154–161). Hershey, PA: IGI Global. doi:10.4018/978-1-59904-947-2.ch013

Seufert, S. (2002). E-learning business models: Framework and best practice examples. In M. Raisinghani (Ed.), *Cases on worldwide e-commerce: Theory in action* (pp. 70–94). Hershey, PA: IGI Global. doi:10.4018/978-1-930708-27-3.ch004

Shareef, M. A., & Archer, N. (2012). E-government service development. In M. Shareef, N. Archer, & S. Dutta (Eds.), *E-government service maturity and development: Cultural, organizational and technological perspectives* (pp. 1–14). Hershey, PA: IGI Global. doi:10.4018/978-1-60960-848-4.ch001

Shareef, M. A., & Archer, N. (2012). E-government initiatives: Review studies on different countries. In M. Shareef, N. Archer, & S. Dutta (Eds.), *E-government service maturity and development: Cultural, organizational and technological perspectives* (pp. 40–76). Hershey, PA: IGI Global. doi:10.4018/978-1-60960-848-4.ch003

Shareef, M. A., Kumar, U., & Kumar, V. (2011). E-government development: Performance evaluation parameters. In M. Shareef, V. Kumar, U. Kumar, & Y. Dwivedi (Eds.), *Stakeholder adoption of e-government services: Driving and resisting factors* (pp. 197–213). Hershey, PA: IGI Global. doi:10.4018/978-1-60960-601-5.ch010

Shareef, M. A., Kumar, U., Kumar, V., & Niktash, M. (2012). Electronic-government vision: Case studies for objectives, strategies, and initiatives. In M. Shareef, N. Archer, & S. Dutta (Eds.), *E-government service maturity and development: Cultural, organizational and technological perspectives* (pp. 15–39). Hershey, PA: IGI Global. doi:10.4018/978-1-60960-848-4.ch002

Shukla, P., Kumar, A., & Anu Kumar, P. B. (2013). Impact of national culture on business continuity management system implementation. *International Journal of Risk and Contingency Management*, *2*(3), 23–36. doi:10.4018/ijrcm.2013070102

Shulman, S. W. (2007). The federal docket management system and the prospect for digital democracy in U S rulemaking. In G. Garson (Ed.), *Modern public information technology systems: Issues and challenges* (pp. 166–184). Hershey, PA: IGI Global. doi:10.4018/978-1-59904-051-6.ch009

Simonovic, S. (2007). Problems of offline government in e-Serbia. In A. Anttiroiko & M. Malkia (Eds.), *Encyclopedia of digital government* (pp. 1342–1351). Hershey, PA: IGI Global. doi:10.4018/978-1-59140-789-8.ch205

Simonovic, S. (2008). Problems of offline government in e-Serbia. In A. Anttiroiko (Ed.), *Electronic government: Concepts, methodologies, tools, and applications* (pp. 2929–2942). Hershey, PA: IGI Global. doi:10.4018/978-1-59904-947-2.ch219

Singh, A. M. (2005). Information systems and technology in South Africa. In M. Khosrow-Pour (Ed.), *Encyclopedia of information science and technology* (pp. 1497–1502). Hershey, PA: IGI Global. doi:10.4018/978-1-59140-553-5.ch263

Singh, S., & Naidoo, G. (2005). Towards an e-government solution: A South African perspective. In W. Huang, K. Siau, & K. Wei (Eds.), *Electronic government strategies and implementation* (pp. 325–353). Hershey, PA: IGI Global. doi:10.4018/978-1-59140-348-7.ch014

Snoke, R., & Underwood, A. (2002). Generic attributes of IS graduates: An analysis of Australian views. In F. Tan (Ed.), *Advanced topics in global information management* (Vol. 1, pp. 370–384). Hershey, PA: IGI Global. doi:10.4018/978-1-930708-43-3.ch023

Sommer, L. (2006). Revealing unseen organizations in higher education: A study framework and application example. In A. Metcalfe (Ed.), *Knowledge management and higher education: A critical analysis* (pp. 115–146). Hershey, PA: IGI Global. doi:10.4018/978-1-59140-509-2.ch007

Song, H., Kidd, T., & Owens, E. (2011). Examining technological disparities and instructional practices in English language arts classroom: Implications for school leadership and teacher training. In L. Tomei (Ed.), *Online courses and ICT in education: Emerging practices and applications* (pp. 258–274). Hershey, PA: IGI Global. doi:10.4018/978-1-60960-150-8.ch020

Speaker, P. J., & Kleist, V. F. (2003). Using information technology to meet electronic commerce and MIS education demands. In A. Aggarwal (Ed.), *Web-based education: Learning from experience* (pp. 280–291). Hershey, PA: IGI Global. doi:10.4018/978-1-59140-102-5.ch017

Spitler, V. K. (2007). Learning to use IT in the workplace: Mechanisms and masters. In M. Mahmood (Ed.), *Contemporary issues in end user computing* (pp. 292–323). Hershey, PA: IGI Global. doi:10.4018/978-1-59140-926-7.ch013

Stellefson, M. (2011). Considerations for marketing distance education courses in health education: Five important questions to examine before development. In U. Demiray & S. Sever (Eds.), *Marketing online education programs: Frameworks for promotion and communication* (pp. 222–234). Hershey, PA: IGI Global. doi:10.4018/978-1-60960-074-7.ch014

Straub, D. W., & Loch, K. D. (2006). Creating and developing a program of global research. *Journal of Global Information Management, 14*(2), 1–28. doi:10.4018/jgim.2006040101

Straub, D. W., Loch, K. D., & Hill, C. E. (2002). Transfer of information technology to the Arab world: A test of cultural influence modeling. In M. Dadashzadeh (Ed.), *Information technology management in developing countries* (pp. 92–134). Hershey, PA: IGI Global. doi:10.4018/978-1-931777-03-2.ch005

Straub, D. W., Loch, K. D., & Hill, C. E. (2003). Transfer of information technology to the Arab world: A test of cultural influence modeling. In F. Tan (Ed.), *Advanced topics in global information management* (Vol. 2, pp. 141–172). Hershey, PA: IGI Global. doi:10.4018/978-1-59140-064-6.ch009

Suki, N. M., Ramayah, T., Ming, M. K., & Suki, N. M. (2013). Factors enhancing employed job seekers intentions to use social networking sites as a job search tool. In A. Mesquita (Ed.), *User perception and influencing factors of technology in everyday life* (pp. 265–281). Hershey, PA: IGI Global. doi:10.4018/978-1-4666-1954-8.ch018

Suomi, R. (2006). Introducing electronic patient records to hospitals: Innovation adoption paths. In T. Spil & R. Schuring (Eds.), *E-health systems diffusion and use: The innovation, the user and the use IT model* (pp. 128–146). Hershey, PA: IGI Global. doi:10.4018/978-1-59140-423-1.ch008

Swim, J., & Barker, L. (2012). Pathways into a gendered occupation: Brazilian women in IT. *International Journal of Social and Organizational Dynamics in IT*, *2*(4), 34–51. doi:10.4018/ijsodit.2012100103

Tarafdar, M., & Vaidya, S. D. (2006). Adoption and implementation of IT in developing nations: Experiences from two public sector enterprises in India. In M. Khosrow-Pour (Ed.), *Cases on information technology planning, design and implementation* (pp. 208–233). Hershey, PA: IGI Global. doi:10.4018/978-1-59904-408-8.ch013

Tarafdar, M., & Vaidya, S. D. (2008). Adoption and implementation of IT in developing nations: Experiences from two public sector enterprises in India. In G. Garson & M. Khosrow-Pour (Eds.), *Handbook of research on public information technology* (pp. 905–924). Hershey, PA: IGI Global. doi:10.4018/978-1-59904-857-4.ch076

Thesing, Z. (2007). Zarina thesing, pumpkin patch. In M. Hunter (Ed.), *Contemporary chief information officers: Management experiences* (pp. 83–94). Hershey, PA: IGI Global. doi:10.4018/978-1-59904-078-3.ch007

Thomas, J. C. (2004). Public involvement in public administration in the information age: Speculations on the effects of technology. In M. Malkia, A. Anttiroiko, & R. Savolainen (Eds.), *eTransformation in governance: New directions in government and politics* (pp. 67–84). Hershey, PA: IGI Global. doi:10.4018/978-1-59140-130-8.ch004

Treiblmaier, H., & Chong, S. (2013). Trust and perceived risk of personal information as antecedents of online information disclosure: Results from three countries. In F. Tan (Ed.), *Global diffusion and adoption of technologies for knowledge and information sharing* (pp. 341–361). Hershey, PA: IGI Global. doi:10.4018/978-1-4666-2142-8.ch015

van Grembergen, W., & de Haes, S. (2008). IT governance in practice: Six case studies. In W. van Grembergen & S. De Haes (Eds.), *Implementing information technology governance: Models, practices and cases* (pp. 125–237). Hershey, PA: IGI Global. doi:10.4018/978-1-59904-924-3.ch004 van Os, G., Homburg, V., & Bekkers, V. (2013). Contingencies and convergence in European social security: ICT coordination in the back office of the welfare state. In M. Cruz-Cunha, I. Miranda, & P. Gonçalves (Eds.), *Handbook of research on ICTs and management systems for improving efficiency in healthcare and social care* (pp. 268–287). Hershey, PA: IGI Global. doi:10.4018/978-1-4666-3990-4.ch013

Velloso, A. B., Gassenferth, W., & Machado, M. A. (2012). Evaluating IBMEC-RJ's intranet usability using fuzzy logic. In M. Cruz-Cunha, P. Gonçalves, N. Lopes, E. Miranda, & G. Putnik (Eds.), *Handbook of research on business social networking: Organizational, managerial, and technological dimensions* (pp. 185–205). Hershey, PA: IGI Global. doi:10.4018/978-1-61350-168-9.ch010

Villablanca, A. C., Baxi, H., & Anderson, K. (2009). Novel data interface for evaluating cardiovascular outcomes in women. In A. Dwivedi (Ed.), *Handbook of research on information technology management and clinical data administration in healthcare* (pp. 34–53). Hershey, PA: IGI Global. doi:10.4018/978-1-60566-356-2.ch003

Villablanca, A. C., Baxi, H., & Anderson, K. (2011). Novel data interface for evaluating cardiovascular outcomes in women. In *Clinical technologies: Concepts, methodologies, tools and applications* (pp. 2094–2113). Hershey, PA: IGI Global. doi:10.4018/978-1-60960-561-2.ch806

Virkar, S. (2011). Information and communication technologies in administrative reform for development: Exploring the case of property tax systems in Karnataka, India. In J. Steyn, J. Van Belle, & E. Mansilla (Eds.), *ICTs for global development and sustainability: Practice and applications* (pp. 127–149). Hershey, PA: IGI Global. doi:10.4018/978-1-61520-997-2.ch006

Virkar, S. (2013). Designing and implementing e-government projects: Actors, influences, and fields of play. In S. Saeed & C. Reddick (Eds.), *Human-centered system design for electronic governance* (pp. 88–110). Hershey, PA: IGI Global. doi:10.4018/978-1-4666-3640-8.ch007

Wallace, A. (2009). E-justice: An Australian perspective. In A. Martínez & P. Abat (Eds.), *E-justice: Using information communication technologies in the court system* (pp. 204–228). Hershey, PA: IGI Global. doi:10.4018/978-1-59904-998-4.ch014

Wang, G. (2012). E-democratic administration and bureaucratic responsiveness: A primary study of bureaucrats' perceptions of the civil service e-mail box in Taiwan. In K. Kloby & M. D'Agostino (Eds.), *Citizen 2.0: Public and governmental interaction through web 2.0 technologies* (pp. 146–173). Hershey, PA: IGI Global. doi:10.4018/978-1-4666-0318-9.ch009

Wangpipatwong, S., Chutimaskul, W., & Papasratorn, B. (2011). Quality enhancing the continued use of e-government web sites: Evidence from e-citizens of Thailand. In V. Weerakkody (Ed.), *Applied technology integration in governmental organizations: New e-government research* (pp. 20–36). Hershey, PA: IGI Global. doi:10.4018/978-1-60960-162-1.ch002

Wedemeijer, L. (2006). Long-term evolution of a conceptual schema at a life insurance company. In M. Khosrow-Pour (Ed.), *Cases on database technologies and applications* (pp. 202–226). Hershey, PA: IGI Global. doi:10.4018/978-1-59904-399-9.ch012

Whybrow, E. (2008). Digital access, ICT fluency, and the economically disadvantages: Approaches to minimize the digital divide. In F. Tan (Ed.), *Global information technologies: Concepts, methodologies, tools, and applications* (pp. 1409–1422). Hershey, PA: IGI Global. doi:10.4018/978-1-59904-939-7.ch102

Whybrow, E. (2008). Digital access, ICT fluency, and the economically disadvantages: Approaches to minimize the digital divide. In C. Van Slyke (Ed.), *Information communication technologies: Concepts, methodologies, tools, and applications* (pp. 764–777). Hershey, PA: IGI Global. doi:10.4018/978-1-59904-949-6.ch049

Wickramasinghe, N., & Geisler, E. (2010). Key considerations for the adoption and implementation of knowledge management in healthcare operations. In M. Saito, N. Wickramasinghe, M. Fuji, & E. Geisler (Eds.), *Redesigning innovative healthcare operation and the role of knowledge management* (pp. 125–142). Hershey, PA: IGI Global. doi:10.4018/978-1-60566-284-8.ch009

Wickramasinghe, N., & Geisler, E. (2012). Key considerations for the adoption and implementation of knowledge management in healthcare operations. In *Organizational learning and knowledge: Concepts, methodologies, tools and applications* (pp. 1316–1328). Hershey, PA: IGI Global. doi:10.4018/978-1-60960-783-8.ch405

Wickramasinghe, N., & Goldberg, S. (2007). A framework for delivering m-health excellence. In L. Al-Hakim (Ed.), *Web mobile-based applications for healthcare management* (pp. 36–61). Hershey, PA: IGI Global. doi:10.4018/978-1-59140-658-7.ch002

Wickramasinghe, N., & Goldberg, S. (2008). Critical success factors for delivering m-health excellence. In N. Wickramasinghe & E. Geisler (Eds.), *Encyclopedia of healthcare information systems* (pp. 339–351). Hershey, PA: IGI Global. doi:10.4018/978-1-59904-889-5.ch045

Wyld, D. (2009). Radio frequency identification (RFID) technology. In J. Symonds, J. Ayoade, & D. Parry (Eds.), *Auto-identification and ubiquitous computing applications* (pp. 279–293). Hershey, PA: IGI Global. doi:10.4018/978-1-60566-298-5.ch017

Yaghmaei, F. (2010). Understanding computerised information systems usage in community health. In J. Rodrigues (Ed.), *Health information systems: Concepts, methodologies, tools, and applications* (pp. 1388–1399). Hershey, PA: IGI Global. doi:10.4018/978-1-60566-988-5.ch088

Yee, G., El-Khatib, K., Korba, L., Patrick, A. S., Song, R., & Xu, Y. (2005). Privacy and trust in e-government. In W. Huang, K. Siau, & K. Wei (Eds.), *Electronic government strategies and implementation* (pp. 145–190). Hershey, PA: IGI Global. doi:10.4018/978-1-59140-348-7.ch007

Yeh, S., & Chu, P. (2010). Evaluation of e-government services: A citizen-centric approach to citizen e-complaint services. In C. Reddick (Ed.), *Citizens and e-government: Evaluating policy and management* (pp. 400–417). Hershey, PA: IGI Global. doi:10.4018/978-1-61520-931-6.ch022

Young-Jin, S., & Seang-tae, K. (2008). E-government concepts, measures, and best practices. In A. Anttiroiko (Ed.), *Electronic government: Concepts, methodologies, tools, and applications* (pp. 32–57). Hershey, PA: IGI Global. doi:10.4018/978-1-59904-947-2.ch004

Yun, H. J., & Opheim, C. (2012). New technology communication in American state governments: The impact on citizen participation. In K. Bwalya & S. Zulu (Eds.), *Handbook of research on e-government in emerging economies: Adoption, e-participation, and legal frameworks* (pp. 573–590). Hershey, PA: IGI Global. doi:10.4018/978-1-4666-0324-0.ch029

Zhang, N., Guo, X., Chen, G., & Chau, P. Y. (2011). User evaluation of e-government systems: A Chinese cultural perspective. In F. Tan (Ed.), *International enterprises and global information technologies: Advancing management practices* (pp. 63–84). Hershey, PA: IGI Global. doi:10.4018/978-1-60960-605-3.ch004

Zuo, Y., & Hu, W. (2011). Trust-based information risk management in a supply chain network. In J. Wang (Ed.), *Supply chain optimization, management and integration: Emerging applications* (pp. 181–196). Hershey, PA: IGI Global. doi:10.4018/978-1-60960-135-5.ch013

Compilation of References

Abdulraheem, M. Y. (1989). Health considerations in using treated industrial and municipal effluents for irrigation. *Desalination DSLNAH*, *72*(1-2), 81–113. doi:10.1016/0011-9164(89)80028-1

Abid, M., Schneider, U. A., & Scheffran, J. (2016). Adaptation to climate change and its impacts on food productivity and crop income: Perspectives of farmers in rural Pakistan. *Journal of Rural Studies*, *47*, 254–266. doi:10.1016/j.jrurstud.2016.08.005

Ackah, I., Alabi, O., & Lartey, A. (2016). Analysing the efficiency of renewable energy consumption among oil-producing African countries. *OPEC Energy Review*, *40*(3), 316–334.

Adhikari, U., Nejadhashemi, A. P., & Woznicki, S. A. (2015). Climate change and eastern Africa: A review of impact on major crops. *Food and Energy Security*, *4*(2), 110–132. doi:10.1002/fes3.61

Adir, Y., Merdler, A., Haim, S. B., Harduf, R., & Bitterman, H. (1999). Effects of exposure to low concentrations of carbon monoxide on exercise performance and myocardial perfusion in young healthy men. *Occupational and Environmental Medicine*, *56*(8), 535–538. doi:10.1136/oem.56.8.535 PMID:10492650

Aggarwal, S. K. (2005). *Water pollution*. Retrieved from https://books.google.co.in/books?id=VRl-24p3ju8C&printsec=frontcover#v=onepage&q&f=false

Aguirre, M., & Ibikunle, G. (2014). Determinants of renewable energy growth: A global sample analysis. *Energy Policy*, *69*, 374–384.

Ahamer, G. (2016). Can educational approaches help to revolutionize quantitative solutions for climate change? In M. Raisinghani (Ed.), *Revolutionizing education through web-based instruction* (pp. 1–19). Hershey, PA: IGI Global. doi:10.4018/978-1-4666-9932-8.ch001

Akhtar, M. N. (2014). Prospective Assessment for Long-Term Impact of Excessive Solid Waste Generation on the Environment. *International Journal of Advancement in Earth and Environmental Sciences*, *2*(2), 39–45.

Akimoto, H. (2017). *Overview of Policy Actions and Observational Data for PM2. 5 and O3 in Japan: A Study of Urban Air Quality Improvement in Asia*. Academic Press.

Akland, G. G., Ott, W. R., & Wallace, L. A. (1984). *Human exposure assessment: Background concepts, purpose, and overview of the Washington, DC.-Denver, Colorado Field Studies. NTIS*.

Alcamo, J., Henrichs, T., & Rösch, T. (2000). *World Water in 2025—Global modeling and scenario analysis for the World Commission on Water for the 21st Century, Kassel World Water Series 2*. Center for Environmental Systems Research, University of Kassel. http://www.usf.uni-kassel.de/usf/archiv/dokumente.en.htm

Allred, E. N., Bleecker, E. R., Chaitman, B. R., Dahms, T. E., Gottlieb, S. O., Hackney, J. D., & Warren, J. et al. (1989). Short-term effects of carbon monoxide exposure on the exercise performance of subjects with coronary artery disease. *The New England Journal of Medicine*, *321*(21), 1426–1432. doi:10.1056/NEJM198911233212102 PMID:2682242

Altman, R., Buckley, W., & Ray, I. (2001). Wet Electrostatic Precipitation Demonstrating Promise for Fine Particulate Control-Part II. *Power Engineering*, *105*(2), 42–42.

Anderson, E. W., Andelman, R. J., Strauch, J. M., Fortuin, N. J., & Knelson, J. H. (1973). Effect of low-level carbon monoxide exposure on onset and duration of angina pectorisa study in ten patients with ischemic heart disease. *Annals of Internal Medicine*, *79*(1), 46–50. doi:10.7326/0003-4819-79-1-46 PMID:4578639

Anderson, T. R., Hawkins, E., & Jones, P. D. (2016). CO_2, the greenhouse effect and global warming: From the pioneering work of Arrhenius and Callendar to today's Earth System Models. *Endeavour*, *40*(3), 178–187. doi:10.1016/j.endeavour.2016.07.002 PMID:27469427

Archer, D., Eby, M., Brovkin, V., Ridgwell, A., Cao, L., & Mikolajewicz, U. et al.. (2009). Atmospheric lifetime of fossil fuel carbon dioxide. *Annual Review of Earth and Planetary Sciences*, *37*(1), 117–134.

Aronow, W. S., Harris, C. N., Isbell, M. W., Rokaw, S. N., & Imparato, B. (1972). Effect of freeway travel on angina pectoris. *Annals of Internal Medicine*, *77*(5), 669–676. doi:10.7326/0003-4819-77-5-669 PMID:4117097

Arora, A., Rao, S. A., Chattopadhyay, R., Goswami, T., George, G., & Sabeerali, C. T. (2016). Role of Indian Ocean SST variability on the recent global warming hiatus. *Global and Planetary Change*, *143*, 21–30. doi:10.1016/j.gloplacha.2016.05.009

Arvidsson, R., & Svanström, M. (2016). A framework for energy use indicators and their reporting in life cycle assessment. *Integrated Environmental Assessment and Management*, *12*(3), 429–436. PMID:26551582

Asolekar & Arceivala. (n.d.). *Wastewater treatment for pollution control and reuse* (3rd ed.). Mc-Graw Hill Education.

ASTM. (2004a). *Standard test method for residual moisture in arefuse-derived fuel analysis sample. E790*. West Conshohocken, PA: ASTM International.

ASTM. (2004b). *Standard test method for ash in the analysis sample ofrefuse-derived fuel. E830*. West Conshohocken, PA: ASTM International.

ASTM. (2004c). *Standard test method for volatile matter in the analysis sample of refuse-derived fuel. E897*. West Conshohocken, PA: ASTM International.

ASTM. (2006a). *Standard test methods for specific gravity of soil solids by water pycnometer. D854*. West Conshohocken, PA: ASTM International.

ASTM. (2008). *Standard test method for determination of the compositionof unprocessed municipal solid waste. D5231-92*. West Conshohocken, PA: ASTM International.

Aune, M., Godbolt, Å. L., Sørensen, K. H., Ryghaug, M., Karlstrøm, H., & Næss, R. (2016). Concerned consumption. Global warming changing household domestication of energy. *Energy Policy*, *98*, 290–297. doi:10.1016/j.enpol.2016.09.001

Austrheim, T., Gjertsen, L. H., & Hoffmann, A. C. (2008). Hoffmann, Alex C. "An experimental investigation of scrubber internals at conditions of low pressure. *Chemical Engineering Journal*, *138*(1), 95–102. doi:10.1016/j.cej.2007.05.048

Ayadi, F. S. (2010). An empirical investigation of environmental Kuznets curve in Nigeria. *International Journal of Green Computing*, *1*(2), 31–39.

Azzopardi, B., & Sanaullah, K. (2002). Re-entrainment in wave-plate mist eliminators. *Chemical Engineering Science*, *57*(17), 3557–3563. doi:10.1016/S0009-2509(02)00270-1

Banik, D. (2016). The Hungry Nation: Food Policy and Food Politics in India. *Food Ethics*, *1*(1), 29–45. doi:10.1007/s41055-016-0001-1

Ban, S. S., Alidina, H. M., Okey, T. A., Gregg, R. M., & Ban, N. C. (2016). Identifying potential marine climate change refugia: A case study in Canada's Pacific marine ecosystems. *Global Ecology and Conservation*, *8*, 41–54. doi:10.1016/j.gecco.2016.07.004

Barth, Monache, Ghude, Pfister, Naja Manish, & Brasseur. (2017). An Overview of Air Quality Modeling Activities in South Asia. In Air Pollution in Eastern Asia: An Integrated Perspective (pp. 27-47). Springer.

Bat, L., & Özkan, E. Y. (2015). Heavy metal levels in sediment of the Turkish Black Sea coast. In I. Zlateva, V. Raykov, & N. Nikolov (Eds.), *Progressive engineering practices in marine resource management* (pp. 399–419). Hershey, PA: IGI Global.

Bayram, H., Devalia, J. L., Khair, O. A., Abdelaziz, M. M., Sapsford, R. J., Sagai, M., & Davies, R. J. (1998). Comparison of ciliary activity and inflammatory mediator release from bronchial epithelial cells of nonatopic non-asthmatic subjects and atopic asthmatic patients and the effect of diesel exhaust particles in vitro. *The Journal of Allergy and Clinical Immunology*, *102*(5), 771–782. doi:10.1016/S0091-6749(98)70017-X PMID:9819294

Bayram, H., Ito, K., & Chung, K. F. (2006). Regulation of human lung epithelial cell numbers by diesel exhaust particles. *The European Respiratory Journal*, *27*(4), 705–713. doi:10.1183/09031936.06.00012805 PMID:16455839

Bayram, H., Khair, O. A., & Abdelaziz, M. (2002). Effect of ozone and nitrogen dioxide on the permeability of bronchial epithelial cell cultures of non-asthmatic and asthmatic subjects. *Clinical and Experimental Allergy*, *32*(9), 1285–1292. doi:10.1046/j.1365-2745.2002.01435.x PMID:12220465

Bayram, H., Sapsford, R. J., Abdelaziz, M., & Khair, O. A. (2001). Effect of ozone and nitrogen dioxide on the release of pro-inflammatory mediators from bronchial epithelial cells of non-atopic non-asthmatic subjects and atopic asthmatic patients, in vitro. *The Journal of Allergy and Clinical Immunology*, *107*(2), 287–294. doi:10.1067/mai.2001.111141 PMID:11174195

Beck, U. (2000). *The brave new world of work*. Cambridge, UK: Polity Press.

Bedient, P. B., Rifai, H. S., & Newell, C. J. (1999). *Ground Water Contamination: Transport and Remediation* (2nd ed.). Prentice Hall Publisher. Retrieved from https://books.google.co.in/books/about/Ground_Water_Contamination.html?id=6RpSAAAAMAAJ&redir_esc=y

Bekele, I., & Ganpat, W. G. (2017). Education, extension, and training for climate change. In *Natural resources management: Concepts, methodologies, tools, and applications* (pp. 279–300). Hershey, PA: IGI Global. doi:10.4018/978-1-5225-0803-8.ch015

Belayutham, S., González, V. A., & Yiu, T. W. (2016). The dynamics of proximal and distal factors in construction site water pollution. *Journal of Cleaner Production, 113*, 54–65.

Benndorf, J. (2008). Ecotechnology and emission control: Alternative or mutually promoting strategies in water resources management? *International Review of Hydrobiology, 93*(4/5), 466–478.

Berend, N. (2016). Contribution of air pollution to COPD and small airway dysfunction. *Respirology (Carlton, Vic.), 21*(2), 237–244. PMID:26412571

Bhagiyalakshmi, M., Anuradha, R., Park, S. D., & Tae, H. J. (2010). Octa (aminophenyl) silsesquioxane fabrication on chlorofunctionalized mesoporous SBA-15 for CO_2 adsorption. *Microporous and Mesoporous Materials, 131*(1-3), 265–273. doi:10.1016/j.micromeso.2010.01.001

Biasioli, M., Fabietti, G., Barberis, R., & Ajmone-Marsan, F. (2012). An appraisal of soil diffuse contamination in an industrial district in northern Italy. *Chemosphere, 88*(10), 1241–1249. PMID:22608707

Bielawski, G. T., & Bhat, P. A. (2001). *Mist elimination/air toxic control in a wet scrubber using a condensing heat exchanger*. U.S. Patent No. 6,273,940. 14 Aug. 2001.

Bircan, H. A. (2003). Effects of atmospheric sulphur dioxide and particulate matter concentrations on emergency room admissions due to asthma in Ankara. *Tüberküloz ve Toraks Dergisi, 51*, 231–238. PMID:15143399

Birmingham, D. P., & Rush, G. C. (2001). *High efficiency gas scrubber using combined coalescing media and centrifugal cyclone*. U.S. Patent No. 6,251,168.

Black, M. (2016). *The Atlas of Water: Mapping the World's Most Critical Resource*. Univ. of California Press.

Blaikie, P., Cannon, T., Davis, I., & Wisner, B. (2014). *At Risk: Natural Hazards, People's Vulnerability and Disasters*. Routledge.

Bogdanović, D., & Lazarević, K. (2017). Early warning system and adaptation advice to reduce human health consequences of extreme weather conditions and air pollution. In *Public health and welfare: Concepts, methodologies, tools, and applications* (pp. 527–564). Hershey, PA: IGI Global.

Borick, C. P., & Rabe, B. G. (2010). A reason to believe: Examining the factors that determine individual views on global warming. *Social Science Quarterly*, *91*(3), 777–800. doi:10.1111/j.1540-6237.2010.00719.x

Bosson, J., Stenfors, N. N., Bucht, A., Helleday, R., Pourazar, J., Holgate, S. T., & Blomberg, A. et al. (2003). Ozone-induced bronchial epithelial cytokine expression differs between healthy and asthmatic subjects. *Clinical and Experimental Allergy*, *33*(6), 777–782. doi:10.1046/j.1365-2222.2003.01662.x PMID:12801312

Boussalis, C., Coan, T. G., & Poberezhskaya, M. (2016). Measuring and modeling Russian newspaper coverage of climate change. *Global Environmental Change*, *41*, 99–110. doi:10.1016/j.gloenvcha.2016.09.004

Boyden, S. A., & Piche, S. (2006). *Control of rolling or moving average values of air pollution control emissions to a desired value*. U.S. Patent No. 7,113,835.

Brown, M. A., & Zhou, S. (2013). Smart-grid policies: An international review. *Wiley Interdisciplinary Reviews: Energy and Environment*, *2*(2), 121–139.

Brown, S. K. (1999). Assessment of Pollutant Emissions from Dry-Process Photocopiers. *Indoor Air*, *9*(4), 259–267. doi:10.1111/j.1600-0668.1999.00005.x PMID:10649859

Brownson, R. C., Baker, E. A., Deshpande, A. D., & Gillespie, K. N. (2017). *Evidence-Based Public Health*. Oxford University Press.

Buckeridge, D. L., Glazier, R., Harvey, B. J., Escobar, M., Amrhein, C., & Frank, J. (2002). Effect of motor vehicle emissions on respiratory health in an urban area. *Environmental Health Perspectives*, *110*(3), 293–300. doi:10.1289/ehp.02110293 PMID:11882481

Buckman, G. (2011). The effectiveness of renewable portfolio standard banding and carve-outs in supporting high-cost types of renewable electricity. *Energy Policy, 39*(7), 4105–4114.

Byun, D., & Schere, K. L. (2006). Review of the governing equations, computational algorithms, and other components of the models-3 community multiscale air quality (CMAQ) modeling system. *Applied Mechanics Reviews, 59*(2), 51–77. doi:10.1115/1.2128636

Cai, Y. P., Huang, G. H., Yeh, S. C., Liu, L., & Li, G. C. (2012). A modeling approach for investigating climate change impacts on renewable energy utilization. *International Journal of Energy Research, 36*(6), 764–777.

Capstick, S., Whitmarsh, L., Poortinga, W., Pidgeon, N., & Upham, P. (2015). International trends in public perceptions of climate change over the past quarter century. *Wiley Interdisciplinary Reviews: Climate Change, 6*(1), 35–61. doi:10.1002/wcc.321

Carvalho, A., Schmidt, L., Santos, F. D., & Delicado, A. (2014). Climate change research and policy in Portugal. *Wiley Interdisciplinary Reviews: Climate Change, 5*(2), 199–217. doi:10.1002/wcc.258

Cavaleri, M. A., Reed, S. C., Smith, W. K., & Wood, T. E. (2015). Urgent need for warming experiments in tropical forests. *Global Change Biology, 21*(6), 2111–2121. doi:10.1111/gcb.12860 PMID:25641092

Census of India. (2011). New Delhi, India: Ministry of Home Affairs, Government of India.

Central Pollution Control Board (CPCB). (2015). *Status of solid waste generation, collection, treatment and disposal in metro cities*. Delhi: Author.

Cernichiari, E., Brewer, R., Myers, G. J., Marsh, D. O., Lapham, L. W., Cox, C., & Clarkson, T. W. et al. (1994). Monitoring methylmercury during pregnancy: Maternal hair predicts fetal brain exposure. *Neurotoxicology, 16*(4), 705–710. PMID:8714874

Cetin, M., Seker, F., & Cavlak, H. (2015). The impact of trade openness on environmental pollution: A panel cointegration and causality analysis. In E. Sorhun, Ü. Hacıoğlu, & H. Dinçer (Eds.), *Regional economic integration and the global financial system* (pp. 221–232). Hershey, PA: IGI Global.

Chakraborty, A., Joshi, P. K., & Sachdeva, K. (2016). Predicting distribution of major forest tree species to potential impacts of climate change in the central Himalayan region. *Ecological Engineering*, *97*, 593–609. doi:10.1016/j.ecoleng.2016.10.006

Chang, T. H., Huang, C. M., & Lee, M. C. (2009). Threshold effect of the economic growth rate on the renewable energy development from a change in energy price: Evidence from OECD countries. *Energy Policy*, *37*(12), 5796–5802.

Chao, C. C., Laffargue, J. P., Liu, X., Sgro, P. M., & Xiao, Y. (2015). Migration and the environment: Policy reform in a polluted open economy. *World Economy*, *38*(1), 48–62.

Chapman, D. (Ed.). (1992). *Water quality assessments. A guide to the use of biota, sediments and water in environmental monitoring*. London: Chapman & Hall. doi:10.4324/9780203476710

Charuvilayil, R. A. (2013). Industrial pollution and people's movement: A case study of Eloor Island Kerala, India. In H. Muga & K. Thomas (Eds.), *Cases on the diffusion and adoption of sustainable development practices* (pp. 312–351). Hershey, PA: IGI Global.

Chaves, I. A., Melchers, R. E., Peng, L., & Stewart, M. G. (2016). Probabilistic remaining life estimation for deteriorating steel marine infrastructure under global warming and nutrient pollution. *Ocean Engineering*, *126*, 129–137. doi:10.1016/j.oceaneng.2016.09.013

Chen, C., Zhang, S., Row, K. H., & Ahn, W.S. (2017). *Amine–silica composites for CO_2 capture: A short review*. Academic Press. 10.1016/j.jechem.2017.07.001

Chen, C., Kim, J., & Ahn, W. S. (2014). CO_2 capture by amine-functionalized nanoporous materials: A review. *Korean Journal of Chemical Engineering*, *31*(11), 1919–1934. doi:10.1007/s11814-014-0257-2

Chen, C., Park, D. W., & Ahn, W. S. (2014). CO_2 capture using zeolite 13X prepared from bentonite. *Applied Surface Science*, *292*, 63–67. doi:10.1016/j.apsusc.2013.11.064

Chen, G., Wan, X., Yang, G., & Zou, X. (2015). Traffic-related air pollution and lung cancer: A meta-analysis. *Thoracic Cancer*, *6*(3), 307–318. PMID:26273377

Ching, J., & Byun, D. (1993). *Introduction to the models-3 framework and the community multiscale air quality model (CMAQ)*. EPA/600/R-99/030.

Cimorelli, A.J. (2004). *AERMOD: Description of model formulation.* EPA-454/R-03-004.

Clayton, C. A., Perritt, R. L., Pellizzari, E. D., Thomas, K. W., Whitmore, R. W., Wallace, L. A., & Spengler, J. D. (1992). Particle Total Exposure Assessment Methodology (PTEAM) study: Distributions of aerosol and elemental concentrations in personal, indoor, and outdoor air samples in a southern California community. *Journal of Exposure Analysis and Environmental Epidemiology*, *3*(2), 227–250. PMID:7694700

Cochard, R. (2011). Consequences of deforestation and climate change on biodiversity. In Y. Trisurat, R. Shrestha, & R. Alkemade (Eds.), *Land use, climate change and biodiversity modeling: Perspectives and applications* (pp. 24–51). Hershey, PA: IGI Global. doi:10.4018/978-1-60960-619-0.ch002

Cortés-Borda, D., Ruiz-Hernández, A., Guillén-Gosálbez, G., Llop, M., Guimerà, R., & Sales-Pardo, M. (2015). Identifying strategies for mitigating the global warming impact of the EU-25 economy using a multi-objective input–output approach. *Energy Policy*, *77*, 21–30. doi:10.1016/j.enpol.2014.11.020

Cosgrove, W. J., & Rijsberman, F. R. (2014). *World Water Vision: Making Water Everybody's Business*. Routledge.

Cowan, N. B., & Abbot, D. S. (2014). Water cycling between ocean and mantle: Super-Earths need not be waterworlds. *The Astrophysical Journal*, *781*(1), 27. doi:10.1088/0004-637X/781/1/27

Da'na, E., & Sayari, A. (2011). Adsorption of copper on amine-functionalized SBA-15 prepared by co-condensation: Equilibrium properties. *Chemical Engineering Journal*, *166*(1), 445–453. doi:10.1016/j.cej.2010.11.016

Danos, A., & Boulouta, K. (2011). The impact of climate change in the modern enterprise. *International Journal of Knowledge Society Research*, *2*(3), 26–35. doi:10.4018/jksr.2011070103

Danzomo, B. A., Salami, M.-J. E., Jibrin, S., & Nor, I. (2012). Performance evaluation of wet scrubber system for industrial air pollution control. *Journal of Engineering and Applied Sciences (Asian Research Publishing Network)*, *7*(12), 1669–1677.

Darunte, L. A., Walton, K. S., Sholl, D. S., & Jones, C. W. (2016). CO_2 capture via adsorption in amine-functionalized sorbents. *Current Opinion in Chemical Engineering*, *12*, 82–90. doi:10.1016/j.coche.2016.03.002

Dasgupta, B., Yadav, V. L., & Mondal, M. K. (2013). Seasonal characterization and present status of municipal solid waste (MSW) management in Varanasi, India. *Advances in Environmental Research*, *2*(1), 51–60. doi:10.12989/aer.2013.2.1.051

Das, R., Ali, E., Hamid, S. B. A., Ramakrishna, S., & Chowdhury, Z. Z. (2014). Carbon nanotube membranes for water purification: A bright future in water desalination. *Desalination*, *336*, 97–109. doi:10.1016/j.desal.2013.12.026

De Asúa, M., & French, R. (2017). *A New World of Animals: Early Modern Europeans on the Creatures of Iberian America*. Routledge.

de Richter, R. K., Ming, T., Caillol, S., & Liu, W. (2016). Fighting global warming by GHG removal: Destroying CFCs and HCFCs in solar-wind power plant hybrids producing renewable energy with no-intermittency. *International Journal of Greenhouse Gas Control*, *49*, 449–472. doi:10.1016/j.ijggc.2016.02.027

de Stefano, M. C., Montes-Sancho, M. J., & Busch, T. (2016). A natural resource-based view of climate change: Innovation challenges in the automobile industry. *Journal of Cleaner Production*, *139*, 1436–1448. doi:10.1016/j.jclepro.2016.08.023

Deguen, S., Ségala, C., Pédrono, G., & Mesbah, M. (2012). A new air quality perception scale for global assessment of air pollution health effects. *Risk Analysis*, *32*(12), 2043–2054. PMID:22852801

del Rio, P., & Burguillo, M. (2009). An empirical analysis of the impact of renewable energy deployment on local sustainability. *Renewable & Sustainable Energy Reviews*, *13*(6/7), 1314–1325.

Delmas, M., & Montes-Sancho, M. (2011). U.S. state policies for renewable energy: Context and effectiveness. *Energy Policy*, *39*(5), 2273–2288.

Desaules, A. (2012). Critical evaluation of soil contamination assessment methods for trace metals. *The Science of the Total Environment*, *426*, 120–131. PMID:22542230

Desmond Ng, J. W., Zhong, Z., Luo, J., & Borgna, A. (2010). Enhancing preferential oxidation of CO in H_2 on Au/α-Fe_2O_3 catalyst via combination with APTES/SBA-15 CO_2-sorbent. *International Journal of Hydrogen Energy*, *35*(23), 12724–12732. doi:10.1016/j.ijhydene.2010.08.136

Dessens, O., Anandarajah, G., & Gambhir, A. (2016). Limiting global warming to 2 °C: What do the latest mitigation studies tell us about costs, technologies and other impacts? *Energy Strategy Reviews*, *13-14*, 67–76. doi:10.1016/j.esr.2016.08.004

Devalia, J. L., Rusznak, C., Herdman, M. J., Trigg, C. J., Davies, R. J., & Tarraf, H. (1994). Effect of nitrogen dioxide and sulphur dioxide on airway response of mild asthmatic patients to allergen inhalation. *Lancet*, *344*(8938), 1668–1671. doi:10.1016/S0140-6736(94)90458-8 PMID:7996960

Devi, J. J., Gupta, T., Jat, R., & Tripathi, S. N. (2013). Measurement of personal and integrated exposure to particulate matter and co-pollutant gases. *Environmental Science and Pollution Research International*, *20*(3), 1632–1648. doi:10.1007/s11356-012-1179-3 PMID:22965544

Devine-Wright, P. (2005). Beyond NIMBYism: Towards an integrated framework for understanding public perceptions of wind energy. *Wind Energy (Chichester, England)*, *8*(2), 125–139.

Devine-Wright, P. (2011). Public engagement with large-scale renewable energy technologies: Breaking the cycle of NIMBYism. *Wiley Interdisciplinary Reviews: Climate Change*, *2*(1), 19–26.

Diamond B. W. Removal of acid mists. (2011). U.S. Patent No. 8,025,860.

Diaz-Sanchez, D. D. (1997). The role of diesel exhaust particles and their associated polyaromatic hydrocarbons in the induction of allergic airway disease. *Allergy*, *52*(suppl 38), 52–56. doi:10.1111/j.1398-9995.1997.tb04871.x PMID:9208060

Didas, S. A., Zhu, R., Brunelli, N. A., Sholl, D. S., & Jones, C. W. (2014). Thermal, oxidative and CO_2 induced degradation of primary amines used for CO_2 capture: Effect of alkyl linker on stability. *The Journal of Physical Chemistry C*, *118*(23), 12302–12311. doi:10.1021/jp5025137

Dittenhofer, M. (1995). Environmental accounting and auditing. *Managerial Auditing Journal*, *10*(8), 40–51. doi:10.1108/02686909510093615

Dockery, D. W. (2001). Epidemiologic evidence of cardiovascular effects of particulate air pollution. *Environmental Health Perspectives*, *109*(Suppl 4), 483–486. doi:10.1289/ehp.01109s4483 PMID:11544151

Dockery, D., Pope, A., Xu, X., Spengler, J., Ware, J., Fay, M., & Speizer, F. et al. (1993). An association between air pollution and mortality in six U.S. cities. *The New England Journal of Medicine*, *329*(24), 1753–1759. doi:10.1056/NEJM199312093292401 PMID:8179653

Dong, C. G. (2012). Feed-in tariff vs. renewable portfolio standard: An empirical test of their relative effectiveness in promoting wind capacity development. *Energy Policy*, *42*, 476–485.

Dong, Z., Pan, Z., Wang, S., An, P., Zhang, J., Zhang, J., & Pan, X. et al. (2016). Effective crop structure adjustment under climate change. *Ecological Indicators*, *69*, 571–577. doi:10.1016/j.ecolind.2016.04.010

dos Santos, M. A. O. (2012). Investigating consumer knowledge of global warming based on Rogers' knowledge stage of the innovation decision process. *International Journal of Consumer Studies*, *36*(4), 385–393. doi:10.1111/j.1470-6431.2011.01069.x

Dresselhaus, M. S., & Tomas, I. L. (2001). Alternative energy technologies. *Nature*, *414*(6861), 332–337. PMID:11713539

Duan, N. (1982). Models for human exposure to air pollution. *Environment International*, *8*(1-6), 305–309. doi:10.1016/0160-4120(82)90041-1

Dudula, J., & Randhir, T. O. (2016). Modeling the influence of climate change on watershed systems: Adaptation through targeted practices. *Journal of Hydrology (Amsterdam)*, *541*, 703–713. doi:10.1016/j.jhydrol.2016.07.020

Elbir, T., Müezzinoğlu, A., & Bayram, A. (2000). Evaluation of some air pollution indicators in Turkey. *Environment International*, *26*(1-2), 5–10. doi:10.1016/S0160-4120(00)00071-4 PMID:11345738

Elijido-Ten, E. O. (2017). Does recognition of climate change related risks and opportunities determine sustainability performance? *Journal of Cleaner Production*, *141*, 956–966. doi:10.1016/j.jclepro.2016.09.136

Elliott, R. J. R., & Shimamoto, K. (2008). Are ASEAN countries havens for Japanese pollution-intensive industry? *World Economy*, *31*(2), 236–254.

Ellis, K. V. (1989). *Surface water pollution and its control*. Retrieved from http://onlinelibrary.wiley.com/doi/10.1002/rrr.3450050408/abstract

Energy Saving Trust. (2013). Retrieved from http://www.energysavingtrust.org.uk/sites/default/files/reports/AtHomewithWater%287%29.pdf

Environmental Protection Agency. (1992). Guidelines for exposure assessment. *Federal Register*, *57*(104), 22887–22938.

Ettler, V. (2016). Soil contamination near non-ferrous metal smelters: A review. *Applied Geochemistry*, *64*, 56–74.

Fagbenro, O. K. (2016). Leachate pollution and impact to environment. In H. Aziz & S. Amr (Eds.), *Control and treatment of landfill leachate for sanitary waste disposal* (pp. 173–199). Hershey, PA: IGI Global.

Fagiani, R., Barquín, J., & Hakvoort, R. (2013). Risk-based assessment of the cost-efficiency and the effectivity of renewable energy support schemes: Certificate markets versus feed-in tariffs. *Energy Policy*, *55*, 648–661.

Fay, J., & Kumar, U. (2013). An index-based model for determining the investment benchmark of renewable energy projects in South Africa. *The South African Journal of Economics*, *81*(3), 416–426.

Finke, T., Gilchrist, A., & Mouzas, S. (2016). Why companies fail to respond to climate change: Collective inaction as an outcome of barriers to interaction. *Industrial Marketing Management*, *58*, 94–101. doi:10.1016/j.indmarman.2016.05.018

Fişekçi, F., Özkurt, S., & Başer, S. (1999). Effect of air pollution on COPD exacerbations. *The European Respiratory Journal*, *14*(Suppl 30), 393s.

Fişekçi, F., Özkurt, S., & Başer, S. (2000). Air pollution and asthma attacks. *The European Respiratory Journal*, *16*(Suppl 31), 290s.

Ford, J. D., Keskitalo, E. C. H., Smith, T., Pearce, T., Berrang-Ford, L., Duerden, F., & Smit, B. (2010). Case study and analogue methodologies in climate change vulnerability research. *Wiley Interdisciplinary Reviews: Climate Change*, *1*(3), 374–392. doi:10.1002/wcc.48

Forino, G., MacKee, J., & von Meding, J. (2016). A proposed assessment index for climate change-related risk for cultural heritage protection in Newcastle (Australia). *International Journal of Disaster Risk Reduction*, *19*, 235–248. doi:10.1016/j.ijdrr.2016.09.003

Fourcade, Y. (2016). Comparing species distributions modelled from occurrence data and from expert-based range maps. Implication for predicting range shifts with climate change. *Ecological Informatics*, *36*, 8–14. doi:10.1016/j.ecoinf.2016.09.002

Frampton, M. W., Boscia, J., Roberts, N. J., Azadniv, M., Torres, A., Cox, C., & Gibb, F. R. et al. (2002). Nitrogen dioxide exposure: Effects on airway and blood cells. *American Journal of Physiology. Lung Cellular and Molecular Physiology*, *282*(1), L155–L165. PMID:11741827

Franchini, M., Guida, A., Tufano, A., & Coppola, A. (2012). Air pollution, vascular disease and thrombosis: Linking clinical data and pathogenic mechanisms. *Journal of Thrombosis and Haemostasis*, *10*(12), 2438–2451. PMID:23006215

Garg, V., & Rani, J. (2012). Perspectives of Municipal Solid Management In India: A Case Study Of Chandigarh. *International Journal of Applied Engineering Research.*

Gazan, R., Sondey, J., Maillot, M., Guelinckx, I., & Lluch, A. (2016). Drinking Water Intake Is Associated with Higher Diet Quality among French Adults. *Nutrients*, *8*(11), 689. doi:10.3390/nu8110689 PMID:27809236

Gezerman, A. O. (2012). A practical solution to abate emission of ammonium nitrate and ammonia during prilling and granulation processes. *International Journal of Medicinal Chemistry*, *2*(2), 57–63.

Gezerman, A. O. (2015). Exergy analysis and purging of an ammonia storage system. *International Journal of Exergy*, *17*(3), 335–351. doi:10.1504/IJEX.2015.070502

Gezerman, A. O. (2016). Industrial-scale purging of ammonia by using nitrogen before environmental discharge. *International Journal of Industrial Chemistry*, *7*(4), 1–8. doi:10.1007/s40090-016-0096-6

Gil, M., Tiscornia, I., Iglesia, Ó., Mallada, R., & Santamaría, J. (2011). Monoamine-grafted MCM-48: An efficient material for CO_2 removal at low partial pressures. *Chemical Engineering Journal*, *175*, 291–297. doi:10.1016/j.cej.2011.09.107

Glaas, E., Ballantyne, A. G., Neset, T. S., & Linnér, B. O. (2017). Visualization for supporting individual climate change adaptation planning: Assessment of a web-based tool. *Landscape and Urban Planning*, *158*, 1–11. doi:10.1016/j.landurbplan.2016.09.018

Gleick, P. (1996). Basic water requirements for human activities: Meeting basic needs. *Water International*, *21*(2), 83–92. doi:10.1080/02508069608686494

Goel, P. K. (2006). *Water Pollution: Causes, Effects and Control.* Retrieved from https://books.google.co.in/books/about/Water_Pollution.html?id=4R9CYYoiFCcC

Goel, S. (2008). Municipal Solid Waste Management (MSWM) in India - A critical Review. *Journal of Environmental Science & Engineering*, *50*(4), 319–328. PMID:19697768

Goklany, I. M. (2012). Is climate change the number one threat to humanity? *Wiley Interdisciplinary Reviews: Climate Change*, *3*(6), 489–508. doi:10.1002/wcc.194

Goldizen, F. C., Sly, P. D., & Knibbs, L. D. (2016). Respiratory effects of air pollution on children. *Pediatric Pulmonology*, *51*(1), 94–108. PMID:26207724

Gomes, A., Pires, J., Figueiredo, S., & Boaventura, R. (2014). Optimization of river water quality surveys by multivariate analysis of physicochemical, bacteriological and ecotoxicological data. *Water Resources Management*, *28*(5), 1345–1361. doi:10.1007/s11269-014-0547-9

Gómez-Sagasti, M. T., Epelde, L., Alkorta, I., & Garbisu, C. (2016). Reflections on soil contamination research from a biologist's point of view. *Applied Soil Ecology*, *105*, 207–210.

Goodman, G. V. R. (2000). Using water sprays to improve performance of a flooded-bed dust scrubber. *Applied Occupational and Environmental Hygiene*, *15*(7), 550–560. doi:10.1080/10473220050028376 PMID:10893791

Goswami, U., & Sharma, H. P. (2007). Study of groundwater contamination due to municipal solid waste dumping in Guwahati city. *Pollution Research*, *26*(2), 211–214.

Goyal, R., & Khare, M. (2010). 4 Indoor Air Pollution and Health Effects. *Health and Environmental Impacts*, 109.

Goyal, M. K., & Royal, I. (2015). Soil carbon sequestration: An alternative option for climate change mitigation. In N. Gaurina-Medjimurec (Ed.), *Handbook of research on advancements in environmental engineering* (pp. 30–54). Hershey, PA: IGI Global. doi:10.4018/978-1-4666-7336-6.ch002

Graham, C. T., & Harrod, C. (2009). Implications of climate change for the fishes of the British Isles. *Journal of Fish Biology*, *74*(6), 1143–1205. doi:10.1111/j.1095-8649.2009.02180.x PMID:20735625

Gray, L. K., Rweyongeza, D., Hamann, A., John, S., & Thomas, B. R. (2016). Developing management strategies for tree improvement programs under climate change: Insights gained from long-term field trials with lodgepole pine. *Forest Ecology and Management*, *377*, 128–138. doi:10.1016/j.foreco.2016.06.041

Grillakis, M. G., Koutroulis, A. G., Seiradakis, K. D., & Tsanis, I. K. (2016). Implications of 2 °C global warming in European summer tourism. *Climate Services*, *1*, 30–38. doi:10.1016/j.cliser.2016.01.002

Gschwend, P. M. (2016). *Environmental Organic Chemistry*. John Wiley & Sons.

Guangming, L., & Zhaofeng, A. (2013). Empirical study on the correlations of environmental pollution, human capital, and economic growth: Based on the 1990-2007 data in Guangdong China. In P. Ordóñez de Pablos (Ed.), *Green technologies and business practices: An IT approach* (pp. 128–137). Hershey, PA: IGI Global.

Guoju, X., Qiang, Z., Fengju, Z., Fei, M., Jing, W., Juying, H., & Zhengji, Q. et al. (2016). Warming influences the yield and water use efficiency of winter wheat in the semiarid regions of Northwest China. *Field Crops Research*, *199*, 129–135. doi:10.1016/j.fcr.2016.09.023

Gutierrez, M. J. (2008). Dynamic inefficiency in an overlapping generation economy with pollution and health costs. *Journal of Public Economic Theory*, *10*(4), 563–594.

Haab, P. (1990). The effect of carbon monoxide on respiration. *Cellular and Molecular Life Sciences*, *46*(11), 1202–1206. doi:10.1007/BF01936937 PMID:2174793

Hadlocon, L. J. S., Manuzon, R. B., & Zhao, L. (2014). Optimization of ammonia absorption using acid spray wet scrubbers. *Transactions of the ASABE*, *57*(2), 647–659.

Hadlocon, L. J. S., Manuzon, R. B., & Zhao, L. (2015). Development and evaluation of a full-scale spray scrubber for ammonia recovery and production of nitrogen fertilizer at poultry facilities. *Environmental Technology*, *36*(4), 405–416. doi:10.1080/09593330.2014.950346 PMID:25518983

Halder, J., & Islam, N. (2015). J.N. Halder1, M.N. Islam, Water Pollution and its Impact on the Human Health. *Journal of Environment and Human*, *2*(1), 36–46. doi:10.15764/EH.2015.01005

Hall, J. E. (2015). *Guyton and Hall Textbook of Medical Physiology E-Book*. Elsevier Health Sciences.

Hammad, A., Akbarnezhad, A., & Rey, D. (2016). Accounting for noise pollution in planning of smart cities. In G. Hua (Ed.), *Smart cities as a solution for reducing urban waste and pollution* (pp. 149–196). Hershey, PA: IGI Global.

Hansen, J., Ruedy, R., Sato, M., & Lo, K. (2010). Global surface temperature change. *Reviews of Geophysics*, *48*(4), RG4004. doi:10.1029/2010RG000345

Haque, M. M., Rahman, A., & Samali, B. (2016). Evaluation of climate change impacts on rainwater harvesting. *Journal of Cleaner Production*, *137*, 60–69. doi:10.1016/j.jclepro.2016.07.038

Hargrove, O. W., & Denlinger, M. A. (2004). *Gas distribution system for venturi scrubbers and absorbers*. U.S. Patent No. 6,808,166.

Harrod, K., Jaramillo, R., Rosenberger, C., Wang, S.-Z., Berger, J. A., McDonald, J. D., & Reed, M. D. (2003). Increased susceptibility to RSV infection by exposure to inhaled diesel engine emissions. *American Journal of Respiratory Cell and Molecular Biology*, *28*(4), 451–463. doi:10.1165/rcmb.2002-0100OC PMID:12654634

Hasselblad, V., Eddy, D. M., & Kotchmar, D. J. (1992). Synthesis of environmental evidence: Nitrogen dioxide epidemiology studies. *Journal of the Air & Waste Management Association*, *42*(5), 662–671. doi:10.1080/10473289.1992.10467018 PMID:1627322

Hazra, T., & Goel, S. (2009). Solid waste management in Kolkata, India: Practices and challenges. *Waste Management (New York, N.Y.)*, *29*(1), 470–478. doi:10.1016/j.wasman.2008.01.023 PMID:18434129

Hedwall, P. O., Skoglund, J., & Linder, S. (2015). Interactions with successional stage and nutrient status determines the life-form-specific effects of increased soil temperature on boreal forest floor vegetation. *Ecology and Evolution*, *5*(4), 948–960. doi:10.1002/ece3.1412 PMID:25750720

Heino, J., Virkkala, R., & Toivonen, H. (2009). Climate change and freshwater biodiversity: Detected patterns, future trends and adaptations in northern regions. *Biological Reviews of the Cambridge Philosophical Society*, *84*(1), 39–54. doi:10.1111/j.1469-185X.2008.00060.x PMID:19032595

Hellmann, F., Alkemade, R., & Knol, O. M. (2016). Dispersal based climate change sensitivity scores for European species. *Ecological Indicators*, *71*, 41–46. doi:10.1016/j.ecolind.2016.06.013

Henschel, D. B., Fortmann, R. C., Roache, N. F., & Liu, X. (2001). Variations in the emissions of volatile organic compounds from the toner for a specific photocopier. *Journal of the Air & Waste Management Association*, *51*(5), 708–717. doi:10.1080/10473289.2001.10464309 PMID:11355458

Hetes, R., Moore, M., & Northeim, C. (1995). Office equipment: design, indoor air emissions, and pollution prevention opportunities. In *Office equipment: design, indoor air emissions, and pollution prevention opportunities*. EPA.

He, Y., Xu, Y., Pang, Y., Tian, H., & Wu, R. (2016). A regulatory policy to promote renewable energy consumption in China: Review and future evolutionary path. *Renewable Energy*, *89*, 695–705.

Heydari-Gorji, A., Yang, Y., & Sayari, A. (2011). Effect of the pore length on CO_2 adsorption over amine-modified mesoporous silicas. *Energy & Fuels*, *25*(9), 4206–4210. doi:10.1021/ef200765f

Heyvaert, V. (2011). Governing climate change: Towards a new paradigm for risk regulation. *The Modern Law Review*, *74*(6), 817–844. doi:10.1111/j.1468-2230.2011.00874.x

Hicks, J. C., Drese, J. H., Fauth, D. J., Gray, M. L., Qi, G., & Jones, C. W. (2008). Designing adsorbents for CO_2 capture from flue gas-hyperbranched aminosilicas capable of capturing CO_2 reversibly. *Journal of the American Chemical Society*, *130*(10), 2902–2903. doi:10.1021/ja077795v PMID:18281986

Ho, C. H., Chen, J. L., Nobuyuki, Y., Lur, H. S., & Lu, H. J. (2016). Mitigating uncertainty and enhancing resilience to climate change in the fisheries sector in Taiwan: Policy implications for food security. *Ocean and Coastal Management*, *130*, 355–372. doi:10.1016/j.ocecoaman.2016.06.020

Hoekstra, A. Y. (2014). Sustainable, efficient, and equitable water use: The three pillars under wise freshwater allocation. *Wiley Interdisciplinary Reviews: Water*, *1*(1), 31–40. doi:10.1002/wat2.1000

Hopkins, D., Campbell-Hunt, C., Carter, L., Higham, J. E. S., & Rosin, C. (2015). Climate change and Aotearoa New Zealand. *Wiley Interdisciplinary Reviews: Climate Change*, *6*(6), 559–583. doi:10.1002/wcc.355

Hori, K., Higuchi, T., Aoki, Y., Miyamoto, M., Yasunori, O., Yogo, K., & Uemiya, S. (2017). Effect of pore size, aminosilane density and aminosilane molecular length on CO_2 adsorption performance in aminosilane modified mesoporous silica. *Microporous and Mesoporous Materials*, *246*, 158–165. doi:10.1016/j.micromeso.2017.03.020

Hosseini, S. E., & Abdul Wahid, M. (2014). The role of renewable and sustainable energy in the energy mix of Malaysia: A review. *International Journal of Energy Research*, *38*(14), 1769–1792.

Hrncevic, L. (2015). Petroleum industry environmental performance and risk. In I. Management Association (Ed.), Transportation systems and engineering: Concepts, methodologies, tools, and applications (pp. 32–56). Hershey, PA: IGI Global.

Hsu, C. Y., Chiang, H. C., Chen, M. J., Chuang, C. Y., Tsen, C. M., Fang, G. C., & Chen, Y. C. et al. (2017). Ambient PM 2.5 in the residential area near industrial complexes: Spatiotemporal variation, source apportionment, and health impact. *The Science of the Total Environment*, *590*, 204–214. doi:10.1016/j.scitotenv.2017.02.212 PMID:28279531

Htun, H., Gray, S. A., Lepczyk, C. A., Titmus, A., & Adams, K. (2016). Combining watershed models and knowledge-based models to predict local-scale impacts of climate change on endangered wildlife. *Environmental Modelling & Software*, *84*, 440–457. doi:10.1016/j.envsoft.2016.07.009

Hua, G. B. (2016). *Smart cities as a solution for reducing urban waste and pollution* (pp. 1–362). Hershey, PA: IGI Global.

Huang, J., & Gurney, K. R. (2016). The variation of climate change impact on building energy consumption to building type and spatiotemporal scale. *Energy*, *111*, 137–153. doi:10.1016/j.energy.2016.05.118

Hua, Q., & Liao, T. F. (2016). The association between rural–urban migration flows and urban air quality in China. *Regional Environmental Change*, *16*(5), 1375–1387. doi:10.1007/s10113-015-0865-3

Huirache-Acuña, R., Nava, R., Peza-Ledesma, C. L., Lara-Romero, J., Alonso-Núñez, G., Pawelec, B., & Rivera-Muñoz, E. M. (2013). SBA-15 Mesoporous silica as catalytic support for hydrodesulfurization catalysts–Review. *Materials (Basel)*, *6*(9), 4139–4167. doi:10.3390/ma6094139 PMID:28788323

Hyde, K. D., Fryar, S., Tian, Q., Bahkali, A. H., & Xu, J. (2016). Lignicolous freshwater fungi along a north–south latitudinal gradient in the Asian/Australian region; can we predict the impact of global warming on biodiversity and function? *Fungal Ecology*, *19*, 190–200. doi:10.1016/j.funeco.2015.07.002

IEHIAS. (n.d.). Retrieved from http://www.integratedassessment.eu/eu/guidebook/indoor_air_pollution_models.html

Indris, A., Inane, B., & Hassan, M. N. (2004). Overview of waste disposal and landfills/dumps in Asian countries. *Journal of Material Cycles and Waste Management*, *6*, 104–110.

Invidiata, A., & Ghisi, E. (2016). Impact of climate change on heating and cooling energy demand in houses in Brazil. *Energy and Building*, *130*, 20–32. doi:10.1016/j.enbuild.2016.07.067

Jaiswal, A., & Chattopadhyaya, M. C. (2015). Water pollution and its treatment. In N. Gaurina-Medjimurec (Ed.), *Handbook of research on advancements in environmental engineering* (pp. 147–171). Hershey, PA: IGI Global.

Jang, Y. S., Lee, D. J., & Oh, H. S. (2013). Evaluation of new and renewable energy technologies in Korea using real options. *International Journal of Energy Research, 37*(13), 1645–1656.

Jennings, A. A., & Li, Z. (2014). Scope of the worldwide effort to regulate pesticide contamination in surface soils. *Journal of Environmental Management, 146*, 420–443. PMID:25199603

Jing, Y., Wei, L., Wang, Y., & Yu, Y. (2014). Synthesis, characterization and CO_2 capture of mesoporous SBA-15 adsorbents functionalized with melamine-based and acrylate-based amine dendrimers. *Microporous and Mesoporous Materials, 183*, 124–133. doi:10.1016/j.micromeso.2013.09.008

Jin, J., Wang, Z., & Ran, S. (2006). Solid waste management in Macao: Practices and challenges. *Waste Management (New York, N.Y.), 26*(9), 1045–1051. doi:10.1016/j.wasman.2005.08.006 PMID:16253497

Jones, S. (2009). The future of renewable energy in Australia: A test for cooperative federalism? *Australian Journal of Public Administration, 68*(1), 1–20.

Jung, H., Lee, C. H., Jeon, S., Jo, D. H., Huh, J., & Kim, S. H. (2016). Effect of amine double-functionalization on CO_2 adsorption behaviors of silica gel-supported adsorbents. *Adsorption, 22*(8), 1137–1146. doi:10.1007/s10450-016-9837-2

Kaarlejärvi, E., Hoset, K. S., & Olofsson, J. (2015). Mammalian herbivores confer resilience of Arctic shrub-dominated ecosystems to changing climate. *Global Change Biology, 21*(9), 3379–3388. doi:10.1111/gcb.12970 PMID:25967156

Kaluarachchi, J. J. (2001). *Groundwater Contamination by Organic Pollutants*. Retrieved from http://ascelibrary.org/doi/book/10.1061/9780784405277

Kanes, D. A., & Wohlgemuth, N. (2008). Evaluation of renewable energy policies in an integrated economic energy environment model. *Forest Policy and Economics, 10*(3), 128–139.

Kanka, H., Jansen, M., Krauss, R., & Kaiser, R. (2012). *U.S. Patent No. 8,328,918*. Washington, DC: U.S. Patent and Trademark Office.

Karlsson, R., & Symons, J. (2015). Making climate leadership meaningful: Energy research as a key to global decarbonisation. *Global Policy*, *6*(2), 107–117.

Kasap, S., Demirer, S. U., & Ergün, S. (2013). An environmentally integrated manufacturing analysis combined with waste management in a car battery manufacturing plant. In I. Management Association (Ed.), Industrial engineering: Concepts, methodologies, tools, and applications (pp. 907–932). Hershey, PA: IGI Global.

Kasemsap, K. (2016a). The roles of lean and green supply chain management strategies in the global business environments. In S. Joshi & R. Joshi (Eds.), *Designing and implementing global supply chain management* (pp. 152–173). Hershey, PA: IGI Global.

Kasemsap, K. (2016b). Multifaceted applications of green supply chain management. In A. Jean-Vasile (Ed.), *Food science, production, and engineering in contemporary economies* (pp. 327–354). Hershey, PA: IGI Global.

Kasemsap, K. (2017a). Environmental management and waste management: Principles and applications. In U. Akkucuk (Ed.), *Ethics and sustainability in global supply chain management* (pp. 26–49). Hershey, PA: IGI Global.

Kasemsap, K. (2017b). Sustainability, environmental sustainability, and sustainable tourism: Advanced issues and implications. In N. Ray (Ed.), *Business infrastructure for sustainability in developing economies* (pp. 1–24). Hershey, PA: IGI Global.

Kasemsap, K. (2017c). Corporate social responsibility: Theory and applications. In A. Bhattacharya (Ed.), *Strategic human capital development and management in emerging economies* (pp. 188–218). Hershey, PA: IGI Global.

Kasemsap, K. (2017d). Mastering consumer attitude and sustainable consumption in the digital age. In N. Suki (Ed.), *Handbook of research on leveraging consumer psychology for effective customer engagement* (pp. 16–41). Hershey, PA: IGI Global.

Kasemsap, K. (2017e). Advocating sustainable supply chain management and sustainability in global supply chain. In M. Khan, M. Hussain, & M. Ajmal (Eds.), *Green supply chain management for sustainable business practice* (pp. 234–271). Hershey, PA: IGI Global.

Katiyar, R. B., Suresh, S., & Sharma, A. K. (2013). Characterization of municipal solid waste generated by city of Bhopal, India, ICGSEE-2013. *Proceedings of International conference on global scenario in Environment and Energy*, *5*(2), 623-628.

Kaygusuz, K., & Sarı, A. (2003). Renewable energy potential and utilization in Turkey. *Energy Conversion and Management*, *44*(3), 459–478. doi:10.1016/S0196-8904(02)00061-4

Keesstra, S., Quinton, J., van der Putten, W., Bardgett, R., & Fresco, L. (2016). The significance of soils and soil science towards realization of the United Nations Sustainable Development Goals. *Soil (Göttingen)*, *2*(2), 111–128. doi:10.5194/soil-2-111-2016

Keles, N., Ilicali, C., & Deger, K. (1999). Impact of air pollution on prevalence of allergic rhinitis in Istanbul. *Archives of Environmental Health*, *54*(1), 48–51. doi:10.1080/00039899909602236 PMID:10025416

Kellstedt, P. M., Zahran, S., & Vedlitz, A. (2008). Personal efficacy, the information environment, and attitudes toward global warming and climate change in the United States. *Risk Analysis*, *28*(1), 113–126. doi:10.1111/j.1539-6924.2008.01010.x PMID:18304110

Khajuria, K., Yamamoto, Y., & Morioka, T. (2010). Estimation of municipal solid waste generation and landfill area in Asian developing countries. *Journal of Environmental Biology*, *31*, 649–654. PMID:21387916

Khatri, R. A., Chuang, S. S. C., Soong, Y., & Gray, M. (2005). Carbon dioxide capture by diamine-grafted sba-15: A combined fourier transform infrared and mass spectrometry study. *Industrial & Engineering Chemistry Research*, *44*(10), 3702–3708. doi:10.1021/ie048997s

Khatun, R., Bhanja, P., Molla, R. A., Ghosh, S., Bhaumik, A., & Islam, S. M. (2017). Functionalized SBA-15 material with grafted CO_2H group as an efficient heterogeneous acid catalyst for the fixation of CO_2 on epoxides under atmospheric pressure. *Molecular Catalysis*, *434*, 25–31. doi:10.1016/j.mcat.2017.01.013

Kilinc-Ata, N. (2016). The evaluation of renewable energy policies across EU countries and US states: An econometric approach. *Energy for Sustainable Development*, *31*, 83–90.

Kim, J., Son, S.-W., Gerber, E., & Park, H.-S. (2017). Defining Sudden Stratospheric Warming in Climate Models: Accounting for Biases in Model Climatologies. *Journal of Climate*, *30*(14), 5529–5546. doi:10.1175/JCLI-D-16-0465.1

Kim, S. B., Temiyasathit, C., Park, S., & Chen, V. C. (2009). Spatio-temporal data mining for air pollution problems. In J. Wang (Ed.), *Encyclopedia of data warehousing and mining* (2nd ed.; pp. 1815–1822). Hershey, PA: IGI Global.

Kim, S. E., Yang, J., & Urpelainen, J. (2016). Does power sector deregulation promote or discourage renewable energy policy? Evidence from the States, 1991–2012. *The Review of Policy Research, 33*(1), 22–50.

Kisić, I. (2015). Effects of soil contamination on the selection of remediation method. In N. Gaurina-Medjimurec (Ed.), *Handbook of research on advancements in environmental engineering* (pp. 200–227). Hershey, PA: IGI Global.

Koenig, J. (2016). 22 Sulfur Dioxide Exposure in Humans. *Toxicology of the Nose and Upper Airways, 334*.

Ko, F. W. S., & Hui, D. S. C. (2012). Air pollution and chronic obstructive pulmonary disease. *Respirology (Carlton, Vic.), 17*(3), 395–401. PMID:22142380

Koike, E., Watanabe, H., & Kobayashi, T. (2004). Exposure to ozone enhances antigen presenting activity concentration dependently in rats. *Toxicology, 197*(1), 37–46. doi:10.1016/j.tox.2003.12.007 PMID:15003332

Kolekar, K. A., Hazra, T., & Chakrabarty, S. N. (2016). A Review on Prediction of Municipal Solid Waste Generation Models. *Procedia Environmental Sciences, 35*, 238–244. doi:10.1016/j.proenv.2016.07.087

Krotoszynski, B. K., Bruneau, G. M., & O'Neill, H. J. (1979). Measurement of chemical inhalation exposure in urban population in the presence of endogenous effluents. *Journal of Analytical Toxicology, 3*(6), 225–234. doi:10.1093/jat/3.6.225

Kuang, Y., Zhang, Y., Zhou, B., Li, C., Cao, Y., Li, L., & Zeng, L. (2016). A review of renewable energy utilization in islands. *Renewable & Sustainable Energy Reviews, 59*, 504–513.

Küçükali, U. F. (2016). Ecological influences on the evolving planning system in Turkey. In U. Benna & S. Garba (Eds.), *Population growth and rapid urbanization in the developing world* (pp. 298–312). Hershey, PA: IGI Global.

Kumar, C. P. (2017). Impact of climate change on groundwater resources. In *Natural resources management: Concepts, methodologies, tools, and applications* (pp. 1094–1120). Hershey, PA: IGI Global. doi:10.4018/978-1-5225-0803-8.ch052

Kumar, P., Geneletti, D., & Nagendra, H. (2016). Spatial assessment of climate change vulnerability at city scale: A study in Bangalore, India. *Land Use Policy, 58*, 514–532. doi:10.1016/j.landusepol.2016.08.018

Kumar, S., Bhattacharyya, J. K., Vaidya, A. N., Chakrabarti, T., Devotta, S., & Akolkar, A. B. (2009). Assessment of the status of municipal solid waste management in metro cities, state capitals, class I cities, and class II towns in India: An insight. *Waste Management (New York, N.Y.)*, *29*(2), 883–895. doi:10.1016/j.wasman.2008.04.011 PMID:18595684

Lakhi, K. S., Cha, W. S., Choy, J. H., Al-Ejji, M., Abdullah, A. M., Al-Enizi, A. M., & Vinu, A. (2016). Synthesis of mesoporous carbons with controlled morphology and pore diameters from SBA-15 prepared through the microwaveassisted process and their CO_2 adsorption capacity. *Microporous and Mesoporous Materials*, *233*, 44–52. doi:10.1016/j.micromeso.2016.06.040

Lang, W., Radke, J. D., Chen, T., & Chan, E. H. W. (2016). Will affordability policy transcend climate change? A new lens to re-examine equitable access to healthcare in the San Francisco Bay Area. *Cities (London, England)*, *58*, 124–136. doi:10.1016/j.cities.2016.05.014

Lawler, J. J. (2009). Climate change adaptation strategies for resource management and conservation planning. *Annals of the New York Academy of Sciences*, *1162*(1), 79–98. doi:10.1111/j.1749-6632.2009.04147.x PMID:19432646

Lawn, J. E., Blencowe, H., Waiswa, P., Amouzou, A., Mathers, C., Hogan, D., & Shiekh, S. (2016). Stillbirths: Rates, risk factors, and acceleration towards 2030. *Lancet*, *387*(10018), 587–603. doi:10.1016/S0140-6736(15)00837-5 PMID:26794078

Lean, H. H., & Smyth, R. (2013). Will policies to promote renewable electricity generation be effective? Evidence from panel stationarity and unit root tests for 115 countries. *Renewable & Sustainable Energy Reviews*, *22*, 371–379.

Leng, G., Huang, M., Tang, Q., & Leung, L. R. (2015). A modeling study of irrigation effects on global surface water and groundwater resources under a changing climate. *Journal of Advances in Modeling Earth Systems*, *7*(3), 1285–1304. doi:10.1002/2015MS000437

Leung, Y. C. D., Caramanna, G., & Maroto-Valer, M. M. (2014). An overview of current status of carbondioxide capture and storage technologies. *Renewable & Sustainable Energy Reviews*, *39*, 426–443. doi:10.1016/j.rser.2014.07.093

Levinson, A., & Taylor, M. S. (2008). Unmasking the pollution haven effect. *International Economic Review*, *49*(1), 223–254.

Liang, J., & Fiorino, D. J. (2013). The implications of policy stability for renewable energy innovation in the United States, 1974–2009. *Policy Studies Journal: the Journal of the Policy Studies Organization*, *41*(1), 97–118.

Li, D., He, J., & Li, L. (2016). A review of renewable energy applications in buildings in the hot-summer and warm-winter region of China. *Renewable & Sustainable Energy Reviews*, *57*, 327–336.

Li, G., Han, H., Du, Y., Hui, D., Xia, J., Niu, S., & Wan, S. et al. (2017). Effects of warming and increased precipitation on net ecosystem productivity: A long-term manipulative experiment in a semiarid grassland. *Agricultural and Forest Meteorology*, *232*, 359–366. doi:10.1016/j.agrformet.2016.09.004

Li, H., & Wang, X. (2012). Study on Chinese low carbon economic model. In D. Ura & P. Ordóñez de Pablos (Eds.), *Advancing technologies for Asian business and economics: Information management developments* (pp. 220–228). Hershey, PA: IGI Global. doi:10.4018/978-1-4666-0276-2.ch017

Li, H., & Zhang, X. (2012). Study on environmental tax: A case of China. In D. Ura & P. Ordóñez de Pablos (Eds.), *Advancing technologies for Asian business and economics: Information management developments* (pp. 207–219). Hershey, PA: IGI Global.

Li, J., Luo, Y., Natali, S., Schuur, E. A. G., Xia, J., Kowalczyk, E., & Wang, Y. (2014). Modeling permafrost thaw and ecosystem carbon cycle under annual and seasonal warming at an Arctic tundra site in Alaska. *Journal of Geophysical Research. Biogeosciences*, *119*(6), 1129–1146. doi:10.1002/2013JG002569

Lin, D., Xia, J., & Wan, S. (2010). Climate warming and biomass accumulation of terrestrial plants: A meta-analysis. *The New Phytologist*, *188*(1), 187–198. doi:10.1111/j.1469-8137.2010.03347.x PMID:20609113

Linfang, W., Lei, M., Aiqin, W., Gian, L., & Tao, Z. (2007). CO_2 Adsorption on SBA-15 Modified by Aminosilane. *Chinese Journal of Catalysis*, *28*(9), 805–810. doi:10.1016/S1872-2067(07)60066-7

Lioy, P. J. (1990). Assessing total human exposure to contaminants. *Environmental Science & Technology*, *24*(7), 938–945. doi:10.1021/es00077a001

Lipfert, F. (1994). *Air Pollution and Community Health- A Critical Review and Data Sourcebook*. Van Nostrand Reinhold.

Li, T., Han, Y., Li, Y., Lu, Z., & Zhao, P. (2016). Urgency, development stage and coordination degree analysis to support differentiation management of water pollution emission control and economic development in the eastern coastal area of China. *Ecological Indicators, 71*, 406–415.

Liu, C., & Wu, X. W. (2011). Factors Influencing Municipal Solid Waste Generation in China: A multiple statistical analysis study. *Waste Management & Research, 29*(4), 371–378. doi:10.1177/0734242X10380114 PMID:20699292

Liu, Y., Lin, X., Wu, X., Liu, M., Shi, R., & Yu, X. (2017). Pentaethylenehexamine loaded SBA-16 for CO_2 capture from simulated flue gas. *Powder Technology, 318*, 186–192. doi:10.1016/j.powtec.2017.06.002

Liu, Y., Mu, J., Niklas, K. J., Li, G., & Sun, S. (2012). Global warming reduces plant reproductive output for temperate multi-inflorescence species on the Tibetan plateau. *The New Phytologist, 195*(2), 427–436. doi:10.1111/j.1469-8137.2012.04178.x PMID:22591333

Livi-Bacci, M. (2001). *A Concise History of World Population* (3rd ed.). Blackwell.

Li, W., Rubin, T. H., & Onyina, P. A. (2013). Comparing solar water heater popularization policies in China, Israel and Australia: The roles of governments in adopting green innovations. *Sustainable Development, 21*(3), 160–170.

Ljungberg, L. Y. (2007). Materials selection and design for development of sustainable products. *Materials & Design, 28*(2), 466–479. doi:10.1016/j.matdes.2005.09.006

Loganathan, S., Tikmani, M., Edubilli, S., Mishra, A., & Ghoshal, A. K. (2014). CO_2 adsorption kinetics on mesoporous silica under wide range of pressure and temperature. *Chemical Engineering Journal, 256*, 1–8. doi:10.1016/j.cej.2014.06.091

Lohmann, H. (2016). Comparing vulnerability and adaptive capacity to climate change in individuals of coastal Dominican Republic. *Ocean and Coastal Management, 132*, 111–119. doi:10.1016/j.ocecoaman.2016.08.009

López-Moreno, J. I., Boike, J., Sanchez-Lorenzo, A., & Pomeroy, J. W. (2016). Impact of climate warming on snow processes in Ny-Ålesund, a polar maritime site at Svalbard. *Global and Planetary Change, 146*, 10–21. doi:10.1016/j.gloplacha.2016.09.006

Louie, A. H., & Pierce, R. C. (1988). Mathematical Models of Human Exposure to air pollutants. *Mathematical and Computer Modelling, 10*(1), 49–64. doi:10.1016/0895-7177(88)90121-5

Lozano, R. (2012). Towards better embedding sustainability into companies' systems: An analysis of voluntary corporate initiatives. *Journal of Cleaner Production*, *25*, 14–26. doi:10.1016/j.jclepro.2011.11.060

Lund, H., Möller, B., Mathiesen, B. V., & Dyrelund, A. (2010). The role of district heating in future renewable energy systems. *Energy*, *35*(3), 1381–1390.

Lund, H., Werner, S., Wiltshire, R., Svendsen, S., Thorsen, J. E., Hvelplund, F., & Mathiesen, B. V. (2014). 4th generation district heating (4GDH): Integrating smart thermal grids into future sustainable energy systems. *Energy*, *68*, 1–11.

Lund, P. (2012). The European Union challenge: Integration of energy, climate, and economic policy. *Wiley Interdisciplinary Reviews: Energy and Environment*, *1*(1), 60–68.

Lu, Y., Song, S., Wang, R., Liu, Z., Meng, J., & Sweetman, A. J. et al.. (2015). Impacts of soil and water pollution on food safety and health risks in China. *Environment International*, *77*, 5–15. PMID:25603422

Maibach, E. W., Kreslake, J. M., Roser-Renouf, C., Rosenthal, S., Feinberg, G., & Leiserowitz, A. A. (2015). Do Americans understand that global warming is harmful to human health? Evidence from a national survey. *Annals of Global Health*, *81*(3), 396–409. doi:10.1016/j.aogh.2015.08.010 PMID:26615074

Maione, M., Fowler, D., Monks, P. S., Reis, S., Rudich, Y., Williams, M. L., & Fuzzi, S. (2016). Air quality and climate change: Designing new win-win policies for Europe. *Environmental Science & Policy*, *65*, 48–57. doi:10.1016/j.envsci.2016.03.011

Mandal, B. K., Ogra, Y., & Suzuki, K. T. (2003). Speciation of arsenic in human nail and hair from arsenic-affected area by HPLC-inductively coupled argon plasma mass spectrometry. *Toxicology and Applied Pharmacology*, *189*(2), 73–83. doi:10.1016/S0041-008X(03)00088-7 PMID:12781625

Mane, T. T., & Hemalata, H. (2012). Existing Situation of Solid Waste Management in Pune City, India. *Research Journal of Recent Sciences, 1*, 348-351.

Manzini, E., Vezzoli, C., & Clark, G. (2001). Product-service systems: Using an existing concept as a new approach to sustainability. *Journal of Desert Research*, *1*(2), 27–40.

Marques, C. A., & Fuinhas, J. A. (2011). Do energy efficiency measures promote the use of renewable sources? *Environmental Science & Policy*, *14*(4), 471–481.

Masters, G. M., & Ela, W. P. (2007). *Introduction to Environmental Engineering and Science*. Retrieved from http://tradownload.co/results/introduction-to-environmental-engineering-and-sciences-by-gilbert-m-and-masters.html

Mathieson, B. V., Henrik Hund, B. V., & Karlson, K. (2011). 100% renewable energy system, climate mitigation and economic growth. *Applied Energy*, *88*(2), 488–501.

Mazumdar, S., & Sussman, N. (1983). Relationships of air pollution to health: Results from the Pittsburgh study. *Archives of Environmental Health: An International Journal*, *38*(1), 17–24. doi:10.1080/00039896.1983.10543974 PMID:6830314

McCright, A. M., Marquart-Pyatt, S. T., Shwom, R. L., Brechin, S. R., & Allen, S. (2016). Ideology, capitalism, and climate: Explaining public views about climate change in the United States. *Energy Research & Social Science*, *21*, 180–189. doi:10.1016/j.erss.2016.08.003

McKenzie, L. M., Witter, R. Z., Newman, L. S., & Adgate, J. L. (2012). Human health risk assessment of air emissions from development of unconventional natural gas resources. *The Science of the Total Environment*, *424*(1), 79–87. PMID:22444058

McKeown, A. E. (2017). Nurses, healthcare, and environmental pollution and solutions: Breaking the cycle of harm. In *Public health and welfare: Concepts, methodologies, tools, and applications* (pp. 110–133). Hershey, PA: IGI Global.

Mekonnen, M. M., & Hoekstra, A. Y. (2016). Four billion people facing severe water scarcity. *Science Advances*, *2*(2), 1–6. doi:10.1126/sciadv.1500323 PMID:26933676

Menegaki, A. N. (2013). Growth and renewable energy in Europe: Benchmarking with data envelopment analysis. *Renewable Energy*, *60*, 363–369.

Meng, Z., Fahong, Z., & Lei, L. (2008). Information technology and environment. In Y. Kurihara, S. Takaya, H. Harui, & H. Kamae (Eds.), *Information technology and economic development* (pp. 201–212). Hershey, PA: IGI Global.

Menyah, K., & Wolde-Rufael, Y. (2010). CO_2 emissions, nuclear energy, renewable energy and economic growth in the US. *Energy Policy*, *38*(6), 2911–2915.

Michelson, I. (1986). *Chevron-type mist eliminator and method.* U.S. Patent No. 4,601,731.

Modic, J. (2003). Carbon monoxide and COHb concentration in blood in various circumstances. *Energy and Building*, *35*(9), 903–907. doi:10.1016/S0378-7788(03)00022-7

Monn, C. (2001). Exposure assessment of air pollutants: A review on spatial heterogeneity and indoor/outdoor/personal exposure to suspended particulate matter, nitrogen dioxide and ozone. *Atmospheric Environment*, *35*(1), 1–32. doi:10.1016/S1352-2310(00)00330-7

Mor, S., Khaiwal, R., Dahiya, R. P., & Chandra, A. (2006b). Municipal Solid Waste Characterization and its Assessment for Potential Methane Generation: A Case Study. *Journal of Science of the Total Environment*, *371*(1-3), 1–10. doi:10.1016/j.scitotenv.2006.04.014 PMID:16822537

Mouratiadou, I., Biewald, A., Pehl, M., Bonsch, M., Baumstark, L., Klein, D., & Kriegler, E. et al. (2016). The impact of climate change mitigation on water demand for energy and food: An integrated analysis based on the shared socioeconomic pathways. *Environmental Science & Policy*, *64*, 48–58. doi:10.1016/j.envsci.2016.06.007

Mukherjee & Nelliyat. (n.d*Groundwater Pollution and Emerging Environmental Challenges of Industrial Effluent Irrigation in Mettupalayam Taluk, Tamil Nadu, Comprehensive Assessment of Water Management in Agriculture*. International Water Management Institute.

Mukherjee, S., & Mukerjee, A. (2017). Sustainable business development by responding to climate change: A case of the Tata group. In *Natural resources management: Concepts, methodologies, tools, and applications* (pp. 416–431). Hershey, PA: IGI Global. doi:10.4018/978-1-5225-0803-8.ch021

Mullan, D., Vandaele, K., Boardman, J., Meneely, J., & Crossley, L. H. (2016). Modelling the effectiveness of grass buffer strips in managing muddy floods under a changing climate. *Geomorphology*, *270*, 102–120. doi:10.1016/j.geomorph.2016.07.012

Myers, D. (1992). Surfactant Science and Technology (2nd ed.). Academic Press.

Nadeau, C. P., & Fuller, A. K. (2016). Combining landscape variables and species traits can improve the utility of climate change vulnerability assessments. *Biological Conservation*, *202*, 30–38. doi:10.1016/j.biocon.2016.07.030

Nadeau, C. P., Fuller, A. K., & Rosenblatt, D. L. (2015). Climate-smart management of biodiversity. *Ecosphere*, *6*(6), 1–17. doi:10.1890/ES15-00069.1

Nanaki, E. A., & Koroneos, C. J. (2016). Climate change mitigation and deployment of electric vehicles in urban areas. *Renewable Energy*, *99*, 1153–1160. doi:10.1016/j.renene.2016.08.006

Nazaroff, W. W. (2004). Indoor particle dynamics. *Indoor Air*, *14*(s7), 175–183. doi:10.1111/j.1600-0668.2004.00286.x PMID:15330785

Nedellec, V., Rabl, A., & Dab, W. (2016). Public health and chronic low chlordecone exposures in Guadeloupe; Part 2: Health impacts, and benefits of prevention. *Environmental Health*, *15*(1), 78. doi:10.1186/s12940-016-0159-3 PMID:27430869

Nemmar, A., Hoet, P. H. M., Dinsdale, D., Vermylen, J., & Hoylaerts, M. F. (2003). Diesel exhaust particles in lung acutely enhance experimental peripheral thrombosis. *Circulation*, *107*(8), 1202–1208. doi:10.1161/01.CIR.0000053568.13058.67 PMID:12615802

Nieuwoudt, I. (2011). *Two-stage mist eliminator and method.* U.S. Patent No. 7,905,937.

Obbard, R., Sadri, S., Wong, Y. Q., Khitun, A. A., Baker, I., & Thompson, R. C. (2014). Global warming releases microplastic legacy frozen in Arctic Sea ice. *Earth's Future*, *2*(6), 315–320. doi:10.1002/2014EF000240

Oliver, T. H., & Morecroft, M. D. (2014). Interactions between climate change and land use change on biodiversity: Attribution problems, risks, and opportunities. *Wiley Interdisciplinary Reviews: Climate Change*, *5*(3), 317–335. doi:10.1002/wcc.271

Ott, W. R., Steinemann, A. C., & Wallace, L. A. (Eds.). (2006). *Exposure analysis.* CRC Press. doi:10.1201/9781420012637.pt1

Øye, G., Sjöblom, J., & Stöcker, M. (2001). Synthesis, characterization and potential applications of new materials in the mesoporous range. *Advances in Colloid and Interface Science*, *89-90*, 439–466. doi:10.1016/S0001-8686(00)00066-X PMID:11215809

Özer, U., Aydin, R., & Akçay, H. (1997). Air pollution profile of Turkey. *Chemistry International*, *19*, 190–191.

Pacesila, M., Burcea, S. G., & Colesca, S. E. (2016). Analysis of renewable energies in European Union. *Renewable & Sustainable Energy Reviews*, *56*, 156–170.

Pan, A., Malik, V. S., Hao, T., Willett, W. C., Mozaffarian, D., & Hu, F. B. (2013). Changes in water and beverage intake and long-term weight changes: Results from three prospective cohort studies. *International Journal of Obesity*, *37*(10), 1378–1385. doi:10.1038/ijo.2012.225 PMID:23318721

Panswad, T., Polprasert, C., & Yamamoto, K. (Eds.). (2016). *Water Pollution Control in Asia: Proceeding of Second IAWPRC Asian Conference on Water Pollution Control Held in Bangkok, Thailand*. Elsevier.

Pappis, C. P. (2011). Global warming: Basic facts. In C. Pappis (Ed.), *Climate change, supply chain management and enterprise adaptation: Implications of global warming on the economy* (pp. 30–65). Hershey, PA: IGI Global. doi:10.4018/978-1-61692-800-1.ch002

Parks, C. R. (2001). *U.S. Patent No. 6,190,438*. Washington, DC: U.S. Patent and Trademark Office.

Pearce, W., Brown, B., Nerlich, B., & Koteyko, N. (2015). Communicating climate change: Conduits, content, and consensus. *Wiley Interdisciplinary Reviews: Climate Change*, *6*(6), 613–626. doi:10.1002/wcc.366

Peavy, H. S., Rowe, D. R., & Tchobanoglous, G. (1985). *Environmental engineering* (International Editions). McGraw-Hill; http://trove.nla.gov.au/work/17834578

Peden, D. (1997). Mechanisms of pollution induced airway disease: In vivo studies. *Allergy*, *52*(suppl 38), 37–44. doi:10.1111/j.1398-9995.1997.tb04869.x PMID:9208058

Peñuelas, J., Sardans, J., Estiarte, M., Ogaya, R., Carnicer, J., Coll, M., & Jump, A. S. et al. (2013). Evidence of current impact of climate change on life: A walk from genes to the biosphere. *Global Change Biology*, *19*(8), 2303–2338. doi:10.1111/gcb.12143 PMID:23505157

Pepper, D. W., & Carrington, D. (2009). *Modeling indoor air pollution*. Imperial College Press. doi:10.1142/p612

Perreault, P., & Shur, Y. (2016). Seasonal thermal insulation to mitigate climate change impacts on foundations in permafrost regions. *Cold Regions Science and Technology*, *132*, 7–18. doi:10.1016/j.coldregions.2016.09.008

Phung, D. T., Hien, T. T., Linh, H. N., Luong, L. M. T., Morawska, L., Chu, C., & Thai, P. K. et al. (2016). Air pollution and risk of respiratory and cardiovascular hospitalizations in the most populous city in Vietnam. *The Science of the Total Environment*, *557*, 322–330. doi:10.1016/j.scitotenv.2016.03.070 PMID:27016680

Pitarma, R. A., Ramos, J. E., Ferreira, M. E., & Carvalho, M. G. (2004). Computational fluid dynamics: An advanced active tool in environmental management and education. *Management of Environmental Quality*, *15*(2), 102–110. doi:10.1108/14777830410523053

Ploberger, C. (2013). A critical assessment of environmental degeneration and climate change: A multidimensional (political, economic, social) challenge for China's future economic development. In Z. Luo (Ed.), *Technological solutions for modern logistics and supply chain management* (pp. 212–229). Hershey, PA: IGI Global. doi:10.4018/978-1-4666-2773-4.ch014

Pokhrel, D., & Viraraghavan, T. (2004). Treatment of pulp and paper mill wastewater—a review. *The Science of the Total Environment*, *333*(1-3), 37–58. doi:10.1016/j.scitotenv.2004.05.017 PMID:15364518

Pollutants, I. (1981). *Committee on Indoor Pollutants*. National Research Council.

Popescu, G. H., & Nica, E. (2017). Global warming, climate policy, and the green paradox. In *Natural resources management: Concepts, methodologies, tools, and applications* (pp. 1–19). Hershey, PA: IGI Global. doi:10.4018/978-1-5225-0803-8.ch001

Popović, V., & Mijajlović, N. (2014). Climate change and sustainable development in agriculture and forestry. In *Sustainable practices: Concepts, methodologies, tools, and applications* (pp. 18–48). Hershey, PA: IGI Global. doi:10.4018/978-1-4666-4852-4.ch002

Popp, D., Hascic, I., & Medhi, N. (2011). Technology and the diffusion of renewable energy. *Energy Economics*, *33*(4), 648–662.

Pratibha, G., Srinivas, I., Rao, K. V., Shanker, A. K., Raju, B. M. K., Choudhary, D. K., & Maheswari, M. et al. (2016). Net global warming potential and greenhouse gas intensity of conventional and conservation agriculture system in rainfed semi arid tropics of India. *Atmospheric Environment*, *145*, 239–250. doi:10.1016/j.atmosenv.2016.09.039

Priac, A., Morin-Crini, N., Druart, C., Gavoille, S., Bradu, C., Lagarrigue, C., & Crini, G. et al. (2017). Alkylphenol and alkylphenolpolyethoxylates in water and wastewater: A review of options for their elimination. *Arabian Journal of Chemistry*, *10*, S3749–S3773. doi:10.1016/j.arabjc.2014.05.011

Pries, C. E. H., van Logtestijn, R. S. P., Schuur, E. A. G., Natali, S. M., Cornelissen, J. H. C., Aerts, R., & Dorrepaal, E. (2015). Decadal warming causes a consistent and persistent shift from heterotrophic to autotrophic respiration in contrasting permafrost ecosystems. *Global Change Biology*, *21*(12), 4508–4519. doi:10.1111/gcb.13032 PMID:26150277

Pueyo, A., & Linares, P. (2012). Renewable technology transfer to developing countries: One size does not fit all. *IDS Working Papers, 2012*(412), 1–39.

Qureshi, M. I., Awan, U., Arshad, Z., Rasli, A. M., Zaman, K., & Khan, F. (2016). Dynamic linkages among energy consumption, air pollution, greenhouse gas emissions and agricultural production in Pakistan: Sustainable agriculture key to policy success. *Natural Hazards*, *84*(1), 367–381. doi:10.1007/s11069-016-2423-9

Rabe, B. G. (2008). States on steroids: The intergovernmental odyssey of American climate policy. *The Review of Policy Research*, *25*(2), 105–128.

Radzi, A. (2015). A survey of expert attitudes on understanding and governing energy autonomy at the local level. *Wiley Interdisciplinary Reviews: Energy and Environment*, *4*(5), 397–405.

Raja, A. A. (2015). *Study of acute toxicity of an unani medicine used in dysentery. Unpublished doctoral dissertation.* BRAC University. Retrieved from http://www.wateraidamerica.org/sites/default/files/attachments/Rainwater%20harvesting.pdf

Ramachandra, T. V., & Varghese, S. K. (2003). Exploring possibilities of achieving sustainability in solid waste management. *Environmental Health*, *45*(4), 255–264. PMID:15527017

Rana, P. R., Yadav, D., Ayub, S., & Siddiqui, A. A. (2014). Status and challenges in solid waste management: A case study of Aligarh City. *Journal of Civil Engineering and Environmental Technology*, *1*(4), 19–24.

Rana, R., Ganguly, R., & Gupta, A. K. (2015). An Assessment of Solid Waste Management System in Chandigarh City, India. *The Electronic Journal of Geotechnical Engineering*, *20*(6), 1547–1572.

Rana, R., Ganguly, R., & Gupta, A. K. (2017). Physico-Chemical Characterization of Municipal Soild Waste from Tri-city region of Northern India – A case study. *Journal of Material Cycles and Waste Management*. doi:10.1007/s10163-017-0615-3

Ranney, M. A., & Clark, D. (2016). Climate change conceptual change: Scientific information can transform attitudes. *Topics in Cognitive Science*, *8*(1), 49–75. doi:10.1111/tops.12187 PMID:26804198

Rao, N. S., Gurunadha Rao, V. V. S., & Gupta, C. P. (1998). Groundwater pollution due to discharge of industrial effluents in Venkatapuram area, Visakhapatnam, Andhra Pradesh, India. *Environmental Geology*, *33*(4), 289–294. doi:10.1007/s002540050248

Rapaczynski, A. (1996). The roles of the state and the market in establishing property rights. *The Journal of Economic Perspectives*, *10*(2), 87–103. doi:10.1257/jep.10.2.87

Rasheed, M. B., Javaid, N., Ahmad, A., Awais, M., Khan, Z. A., Qasim, U., & Alrajeh, N. (2016). Priority and delay constrained demand side management in real-time price environment with renewable energy source. *International Journal of Energy Research*, *40*(14), 2002–2021.

Rathoure, A. K. (2016). Heavy metal pollution and its management: Bioremediation of heavy metal. In A. Rathoure & V. Dhatwalia (Eds.), *Toxicity and waste management using bioremediation* (pp. 27–50). Hershey, PA: IGI Global.

Rawat, M., Ramanathan, A. L., & Kuriakose, T. (2013). Characterization of municipal solid waste compost from selected from selected Indian cities- A case study for its sustainable utilization. *Environment and Progress*, *4*(2), 163–171. doi:10.4236/jep.2013.42019

Recalde, M. Y. (2016). The different paths for renewable energies in Latin American countries: The relevance of the enabling frameworks and the design of instruments. *Wiley Interdisciplinary Reviews: Energy and Environment*, *5*(3), 305–326.

Rene, E., Jarrier, D. J., & Bryant, P. (2012). *U.S. Patent No. 8,273,158*. Washington, DC: U.S. Patent and Trademark Office.

Reynolds, T. W., Bostrom, A., Read, D., & Morgan, M. G. (2010). Now what do people know about global climate change? Survey studies of educated laypeople. *Risk Analysis*, *30*(10), 1520–1538. doi:10.1111/j.1539-6924.2010.01448.x PMID:20649942

Rocco, C., Duro, I., di Rosa, S., Fagnano, M., Fiorentino, N., Vetromile, A., & Adamo, P. (2016). Composite vs. discrete soil sampling in assessing soil pollution of agricultural sites affected by solid waste disposal. *Journal of Geochemical Exploration*, *170*, 30–38.

Rousse, O. (2008). Environmental and economic benefits resulting from citizens' participation in CO_2 emissions trading: An efficient alternative solution to the voluntary compensation of CO_2 emissions. *Energy Policy*, *36*(1), 388–397. doi:10.1016/j.enpol.2007.09.019

Roy, S. D. A., & Bandyopadhyay, S. (2011). Testbed implementation of a pollution monitoring system using wireless sensor network for the protection of public spaces. In V. Sridhar & D. Saha (Eds.), *Recent advances in broadband integrated network operations and services management* (pp. 263–276). Hershey, PA: IGI Global.

Rubio-Bellido, C., Pérez-Fargallo, A., & Pulido-Arcas, J. A. (2016). Optimization of annual energy demand in office buildings under the influence of climate change in Chile. *Energy*, *114*, 569–585. doi:10.1016/j.energy.2016.08.021

Rusznak, C., Devalia, J. L., Bayram, H., & Davies, R. J. (1997). Impact of the environment on allergic lung diseases. *Clinical and Experimental Allergy*, *27*(s1), 26–35. doi:10.1111/j.1365-2222.1997.tb01823.x PMID:9179442

Safi, A. S., Smith, W. J. Jr, & Liu, Z. (2012). Rural Nevada and climate change: Vulnerability, beliefs, and risk perception. *Risk Analysis*, *32*(6), 1041–1059. doi:10.1111/j.1539-6924.2012.01836.x PMID:22583075

Saha, J. K., Panwar, N., & Singh, M. V. (2010). An assessment of municipal solid waste compost quality produced in different cities of India in the perspective of developing quality control indices. *Waste Management (New York, N.Y.)*, *30*(2), 192–201. doi:10.1016/j.wasman.2009.09.041 PMID:19857948

Saksena, S. (2011). Public perceptions of urban air pollution risks. *Risk, Hazards & Crisis in Public Policy*, *2*(1), 1–19.

Sallam, G. A., Youssef, T., Embaby, M. E., & Shaltot, F. (2011). Using geographic information system to infollow the fertilizers pollution migration. In *Green technologies: Concepts, methodologies, tools and applications* (pp. 564–586). Hershey, PA: IGI Global.

Samiey, B., Cheng, C. H., & Wu, J. (2014). Organic-inorganic hybrid polymers as adsorbents for removal of heavy metal ions from solutions: A Review. *Materials (Basel)*, *7*(2), 673–726. doi:10.3390/ma7020673 PMID:28788483

Sango, H. A., Testa, J., Meda, N., Contrand, B., Traoré, M. S., Staccini, P., & Lagarde, E. (2016). Mortality and morbidity of urban road traffic crashes in Africa: Capture-recapture estimates in Bamako, Mali. *PLoS One*, *11*(2), e0149070. doi:10.1371/journal.pone.0149070 PMID:26871569

Santamouris, M., Cartalis, C., Synnefa, A., & Kolokotsa, D. (2015). On the impact of urban heat island and global warming on the power demand and electricity consumption of buildings: A review. *Energy and Building*, *98*, 119–124. doi:10.1016/j.enbuild.2014.09.052

Sanz-Pérez, E. S., Murdock, C. R., Didas, S. A., & Jones, C. W. (2016). Direct capture of CO_2 from ambient air. *Chemical Reviews*, *116*(19), 11840–11876. doi:10.1021/acs.chemrev.6b00173 PMID:27560307

Sanz-Pérez, E. S., Olivares-Marín, M., Arencibia, A., Sanz, R., Calleja, G., & Maroto-Valer, M. M. (2013). CO_2 adsorption performance of amino-functionalized SBA-15 under post-combustion conditions. *International Journal of Greenhouse Gas Control, 17*, 366–375. doi:10.1016/j.ijggc.2013.05.011

Sanz, R., Calleja, G., Arencibia, A., & Sanz-Pérez, E. S. (2012). Amino functionalized mesostructured SBA-15 silica for CO_2 capture: Exploring the relation between the adsorption capacity and the distribution of amino groups by TEM. *Microporous and Mesoporous Materials, 158*, 309–317. doi:10.1016/j.micromeso.2012.03.053

Sardianou, E., & Genoudi, P. (2013). Which factors affect the willingness of consumers to adopt renewable energies? *Renewable Energy, 57*, 1–4.

Sarkar, A. N. (2013). Promoting eco-innovations to leverage sustainable development of eco-industry and green growth. *European Journal of Sustainable Development, 2*(1), 171.

Sastry, N. (2002). Forest fires, air pollution, and mortality in southeast Asia. *Demography, 39*(1), 1–23. doi:10.1353/dem.2002.0009 PMID:11852832

Schaefer, K., Zhang, T., Bruhwiler, L., & Barrett, A. P. (2011). Amount and timing of permafrost carbon release in response to climate warming. *Tellus. Series B, Chemical and Physical Meteorology, 63*(2), 165–180. doi:10.1111/j.1600-0889.2011.00527.x

Schliephack, J., & Dickinson, J. E. (2017). Tourists' representations of coastal managed realignment as a climate change adaptation strategy. *Tourism Management, 59*, 182–192. doi:10.1016/j.tourman.2016.08.004

Schmeida, M., & McNeal, R. S. (2017). U.S. public support to climate change initiatives?: Setting stricter carbon dioxide emission limits on power plants. In *Natural resources management: Concepts, methodologies, tools, and applications* (pp. 1196–1215). Hershey, PA: IGI Global. doi:10.4018/978-1-5225-0803-8.ch058

Seo, S. N. (2013). Economics of global warming as a global public good: Private incentives and smart adaptations. *Regional Science Policy & Practice, 5*(1), 83–95. doi:10.1111/j.1757-7802.2012.01088.x

Setekleiv, E., & Svendsen, H. (2010, January). Scrubber performance: Investigation of liquid holdup in mesh pad at ambient conditions with applications to high pressure separation. *Offshore Technology Conference*. doi:10.4043/20390-MS

Sethi, S., Kothiyal, N. C., Nema, A. K., & Kaushik, M. K. (2013). Characterization of Municipal Solid Waste in Jalandhar City, Punjab, India. *Journal of Hazardous, Toxic and Radioactive Waste*, *17*(2), 97–106. doi:10.1061/(ASCE)HZ.2153-5515.0000156

Shah, P., Ramaswamy, A. V., Lazar, K., & Ramaswamy, V. (2007). Direct hydrothermal synthesis of mesoporous Sn-SBA-15 materials under weak acidic conditions. *Microporous and Mesoporous Materials*, *100*(1-3), 210–226. doi:10.1016/j.micromeso.2006.10.042

Shakerian, F., Kim, K. H., Szulejko, J. E., & Park, J. W. (2015). A comparative review between amines and ammonia as sorptive media for post-combustion CO_2 capture. *Applied Energy*, *148*, 10–22. doi:10.1016/j.apenergy.2015.03.026

Sharholy, M., Ahmad, K., Mahmood, G., & Trivedi, R. C. (2008). Municipal solid waste management in Indian cities - A review. *Waste Management (New York, N.Y.)*, *28*(2), 459–467. doi:10.1016/j.wasman.2007.02.008 PMID:17433664

Sharma, Ojha, Pozzer, Mar, Beig, Lelieveld, & Gunthe. (2016). *WRF-Chem simulated surface ozone over South Asia during the pre-monsoon: Effects of emission inventories and chemical 2 mechanisms 3*. Academic Press.

Sharma, A., Kaur, M., Katnoria, J. K., & Nagpal, A. K. (2016). Heavy metal pollution: A global pollutant of rising concern. In A. Rathoure & V. Dhatwalia (Eds.), *Toxicity and waste management using bioremediation* (pp. 1–26). Hershey, PA: IGI Global.

Sharma, S. (2016). Economic damage due to ozone pollution in NCR: Ozone impacts. In A. Goswami & A. Mishra (Eds.), *Economic modeling, analysis, and policy for sustainability* (pp. 284–306). Hershey, PA: IGI Global.

Sheikhavandi, T. (2015). Soil contamination. In N. Gaurina-Medjimurec (Ed.), *Handbook of research on advancements in environmental engineering* (pp. 173–199). Hershey, PA: IGI Global.

Shi, D. (2009). Analysis of China's renewable energy development under the current economic and technical circumstances. *China & World Economy*, *17*(2), 94–109.

Short, F. T., Kosten, S., Morgan, P. A., Malone, S., & Moore, G. E. (2016). Impacts of climate change on submerged and emergent wetland plants. *Aquatic Botany*, *135*, 3–17. doi:10.1016/j.aquabot.2016.06.006

Sikkema, R., Steiner, M., Junginger, M., Hiegl, W., Hansen, M. T., & Faaij, A. (2011). The European wood pellet markets: Current status and prospects for 2020. *Biofuels, Bioproducts & Biorefining*, *5*(3), 250–278.

Singh, K. P., & Kushwaha, C. P. (2016). Deciduousness in tropical trees and its potential as indicator of climate change: A review. *Ecological Indicators*, *69*, 699–706. doi:10.1016/j.ecolind.2016.04.011

Singh, K. P., & Parwana, H. K. (1998). Groundwater pollution due to industrial wastewater in Punjab state and strategies for its control. *Indian Journal of Environmental Protection*, *19*, 241–244.

Skamp, K., Boyes, E., & Stanisstreet, M. (2013). Beliefs and willingness to act about global warming: Where to focus science pedagogy? *Science Education*, *97*(2), 191–217. doi:10.1002/sce.21050

Snyder, P., Lockett, M., Moustakas, D., & Whitesides, G. (2014). Is it the shape of the cavity, or the shape of the water in the cavity? *The European Physical Journal. Special Topics*, *223*(5), 853–891. doi:10.1140/epjst/e2013-01818-y

Sorensen, I. W., & Lamb, E. A. (2000). *U.S. Patent No. 6,068,686*. Washington, DC: U.S. Patent and Trademark Office.

Specht, E., Redemann, T., & Lorenz, N. (2016). Simplified mathematical model for calculating global warming through anthropogenic CO_2. *International Journal of Thermal Sciences*, *102*, 1–8. doi:10.1016/j.ijthermalsci.2015.10.039

Spellman, F. R. (2017). *The Drinking Water Handbook*. CRC Press. doi:10.1201/9781315159126

Spengler, J. D., Samet, J. M., & McCarthy, J. F. (2001). *Indoor air quality handbook*. New York: McGraw-Hill.

Spengler, J. D., & Sexton, K. (1983). Indoor air pollution—a public health perspective. *Science*, *221*(4605), 9–17. doi:10.1126/science.6857273 PMID:6857273

Spengler, J. D., & Wilson, R. (1996). *Particles in Our Air, Concentrations and Health Effects*. Harvard University Press.

Srivastava, N. (2017). Climate change mitigation: Collective efforts and responsibly. In *Natural resources management: Concepts, methodologies, tools, and applications* (pp. 64–76). Hershey, PA: IGI Global. doi:10.4018/978-1-5225-0803-8.ch003

Stenfors, N., Nordenhall, C., Salvi, S., Mudway, I., Soderberg, M., Blomberg, A., & Sandstrom, T. et al. (2004). Different airway inflammatory responses in asthmatic and healthy humans exposed to diesel. *The European Respiratory Journal*, *23*(1), 82–86. doi:10.1183/09031936.03.00004603 PMID:14738236

Sterk, A., de Man, H., Schijven, J. F., de Nijs, T., & de Roda Husman, A. M. (2016). Climate change impact on infection risks during bathing downstream of sewage emissions from CSOs or WWTPs. *Water Research*, *105*, 11–21. doi:10.1016/j.watres.2016.08.053 PMID:27591704

Stigka, E. K., Paravantis, J. A., & Mihalakakou, G. K. (2014). Social acceptance of renewable energy sources: A review of contingent valuation applications. *Renewable & Sustainable Energy Reviews*, *32*, 100–106.

Stokes, C. L. (2013). The politics of renewable energy policies: The case of feed-in tariffs in Ontario, Canada. *Energy Policy*, *56*, 490–500.

Subrahmanyam, K., & Yadaiah, P. (2001). Assessment of the impact of industrial effluents on water quality in Patancheru and environs, Medak district, Andhra Pradesh, India. *Hydrogeology Journal*, *9*(3), 297–312. doi:10.1007/s100400000120

Suchak, N., Finley, S. J., Eschbach, J. A., & Aeiss, R. (2009). *U.S. Patent No. 7,632,475*. Washington, DC: U.S. Patent and Trademark Office.

Sun, P., & Yuan, Y. (2015). Industrial agglomeration and environmental degradation: Empirical evidence in Chinese cities. *Pacific Economic Review*, *20*(4), 544–568.

Talyan, V., Dahiya, R. P., & Sreekrishnan, T. R. (2008). State of municipal solid waste management in Delhi, the capital of India. *Waste Management (New York, N.Y.)*, *28*(7), 1276–1287. doi:10.1016/j.wasman.2007.05.017 PMID:17692510

Tan, K. (2014). *Humic Matter in Soil and the Environment: Principles and Controversies*. CRC Press, Taylor & Francis Group.

Tang, C., Yi, Y., Yang, Z., & Sun, J. (2016). Risk analysis of emergent water pollution accidents based on a Bayesian network. *Journal of Environmental Management*, *165*, 199–205. PMID:26433361

Taylan, V., Dahiya, R. P., & Anand, S. (2007). Quantification of Methane emission from Solid Waste Disposal in Delhi. *Journal of Resources Conservation and Recycling*, *3*(3), 240–259. doi:10.1016/j.resconrec.2006.06.002

Tchobaanoglous, H., Theisen, H., & Samuel, A. (1993). *Integrated Solid Waste Management*. New Delhi: McGraw-Hill, Inc.

Thakur, I., Ghosh, P., & Gupta, A. (2015). Combined chemical and toxicological evaluation of leachate from municipal solid waste landfill sites of Delhi, India. *Environmental Science and Pollution Research International*, *22*(12), 9148–9158. doi:10.1007/s11356-015-4077-7 PMID:25578612

The Gazette of India. (2000). *Municipal solid waste (management and handling) rules, Ministry of Environment and Forests (MoEF)*. Author.

Thunyaratchatanon, C., Luengnaruemitchai, A., Chaisuwan, T., Chollacoop, N., Chen, S. Y., & Yoshimura, Y. (2017). Synthesis and characterization of Zr incorporation into highly ordered mesostructured SBA-15 material and its performance for CO_2 adsorption. *Microporous and Mesoporous Materials*, *253*, 18–28. doi:10.1016/j.micromeso.2017.06.015

Tong, H. (2016). Dietary and pharmacological intervention to mitigate the cardiopulmonary effects of air pollution toxicity. *Biochimica et Biophysica Acta (BBA) -. General Subjects*, *1860*(12), 2891–2898. PMID:27189803

Tran, N. H., Gin, K. Y. H., & Ngo, H. H. (2015). Fecal pollution source tracking toolbox for identification, evaluation and characterization of fecal contamination in receiving urban surface waters and groundwater. *The Science of the Total Environment*, *538*, 38–57. PMID:26298247

Tricys, V. (2002). Research of leachate, surface and ground water pollution near Siauliai landfill. *Environmental Research, Engineering and Management*, *19*, 30–33.

Tütüncü, S. İ., & Şahin, N. (2012). Üretimde Kirlilik Önleme Yaklaşımı ve Devletin Konumu. *Ekonomi Bilimleri Dergisi*, *4*(2).

Ullah, R., Atilhan, M., Aparicio, S., Canlier, A., & Yavuz, C. T. (2015). Insights of CO_2 adsorption performance of amine impregnated mesoporous silica (SBA-15) at wide range pressure and temperature conditions. *International Journal of Greenhouse Gas Control*, *43*, 22–32. doi:10.1016/j.ijggc.2015.09.013

Unveren, E. E., Monkul, B. O., Sarıoğlan, Ş., Karademir, N., & Alper, E. (2017). Solid amine sorbents for CO_2 capture by chemical adsorption: A review. *Petroleum*, *3*(1), 37–50. doi:10.1016/j.petlm.2016.11.001

Urban, F. (2009). Climate-change mitigation revisited: Low-carbon energy transitions for China and India. *Development Policy Review*, *27*(6), 693–715.

Uyterlinde, M. A., Junginger, M., de Vries, H. J., Faaij, A. P. C., & Turkenburg, W. C. (2007). Implications of technological learning on the prospects for renewable energy technologies in Europe. *Energy Policy*, *35*(8), 4072–4087.

VanBuskirk, G. K., Hsieh, C.-L., McNulty, K. J., & Hansen, O. V. (1995). *U.S. Patent No. 5,464,459*. Washington, DC: U.S. Patent and Trademark Office.

Vedal, S. (1997). Critical review-Ambient particles and health: Lines that divide. *Journal of the Air & Waste Management Association*, *47*(5), 551–581. doi:10.1080/10473289.1997.10463922 PMID:9155246

Voulvoulis, N., & Georges, K. (2016). Industrial and agricultural sources and pathways of aquatic pollution. In A. McKeown & G. Bugyi (Eds.), *Impact of water pollution on human health and environmental sustainability* (pp. 29–54). Hershey, PA: IGI Global.

Walck, J. L., Hidayati, S. N., Dixon, K. W., Thompson, K., & Poschlod, P. (2015). Climate change and plant regeneration from seed. *Global Change Biology*, *17*(6), 2145–2161. doi:10.1111/j.1365-2486.2010.02368.x

Wang, C., Cai, J., Chen, R., Shi, J., Yang, C., Li, H., & Xia, Y. et al. (2017). Personal exposure to fine particulate matter, lung function and serum club cell secretory protein (Clara). *Environmental Pollution*, *225*, 450–455. doi:10.1016/j.envpol.2017.02.068 PMID:28284549

Wang, C., & Dong, S. (2010). Is the basin-wide warming in the North Atlantic Ocean related to atmospheric carbon dioxide and global warming? *Geophysical Research Letters*, *37*(8), L08707. doi:10.1029/2010GL042743

Wang, H. F., Sung, M. P., & Hsu, H. W. (2016). Complementarity and substitution of renewable energy in target year energy supply-mix plannin–in the case of Taiwan. *Energy Policy*, *90*, 172–182.

Wang, I. J., Karmaus, W. J., & Yang, C. C. (2017). Lead exposure, IgE, and the risk of asthma in children. *Journal of Exposure Science & Environmental Epidemiology*, *27*(5), 478–483. doi:10.1038/jes.2017.5 PMID:28401896

Wang, Q., & Yang, Z. (2016). Industrial water pollution, water environment treatment, and health risks in China. *Environmental Pollution*, *218*, 358–365. PMID:27443951

Wang, X., & Song, C. (2012). Temperature-programmed desorption of CO_2 from polyethylenimine-loaded SBA-15 as molecular basket sorbents. *Catalysis Today*, *194*(1), 44–52. doi:10.1016/j.cattod.2012.08.008

Wang, X., & Song, C. (2014). New strategy to enhance CO_2 capture over a nanoporous polyethylenimine sorbent. *Energy & Fuels*, *28*(12), 7742–7745. doi:10.1021/ef501997q

Wayne, R. O. (1982). Concepts of human exposure to air pollution. *Environment International*, *7*(3), 179–196. doi:10.1016/0160-4120(82)90104-0

Wei, J., Shi, J., Pan, H., Zhao, W., Ye, Q., & Shi, Y. (2008). Adsorption of carbon dioxide on organically functionalized SBA-16. *Microporous and Mesoporous Materials*, *116*(1-3), 394–399. doi:10.1016/j.micromeso.2008.04.028

Weil, R. R., & Brady, N. C. (2016). *The Nature and Properties of Soils*. Columbus, OH: Pearson.

Weng, Y. C., Fujiwara, T., Houng, H. J., Sun, C. H., Li, W. Y., & Kuo, Y. W. (2015). Management of landfill reclamation with regard to biodiversity preservation, global warming mitigation and landfill mining: Experiences from the Asia–Pacific region. *Journal of Cleaner Production*, *104*(1), 364–373. doi:10.1016/j.jclepro.2015.05.014

Wesley, B., Anderson, D. K., & Kingston, W. H. (2000). *U.S. Patent No. 6,083,302*. Washington, DC: U.S. Patent and Trademark Office.

West, J., Cohen, A., Dentener, F., Brunekreef, B., Zhu, T., Armstrong, B., & Wiedinmyer, C. et al. (2016). What we breathe impacts our health: Improving understanding of the link between air pollution and health. *Environmental Science & Technology*, *50*(10), 4895–4904. doi:10.1021/acs.est.5b03827 PMID:27010639

WHO-United Nations Environment Programme. (1992). *Urban Air Pollution in Megacities of the World*. Blackwell.

Wiese, F., Bökenkamp, G., Wingenbach, C., & Hohmeyer, O. (2014). An open source energy system simulation model as an instrument for public participation in the development of strategies for a sustainable future. *Wiley Interdisciplinary Reviews: Energy and Environment*, *3*(5), 490–504.

Wilson, P. (2015). Farm-level actions towards water pollution control: The role of nutrient guidance systems. *Water and Environment Journal: the Journal / the Chartered Institution of Water and Environmental Management*, *29*(1), 88–97.

Winter, R. A. (2014). Innovation and the dynamics of global warming. *Journal of Environmental Economics and Management*, *68*(1), 124–140. doi:10.1016/j.jeem.2014.01.005

Wolf, J., & Moser, S. C. (2011). Individual understandings, perceptions, and engagement with climate change: Insights from in-depth studies across the world. *Wiley Interdisciplinary Reviews: Climate Change*, *2*(4), 547–569. doi:10.1002/wcc.120

Yanagisawa, Y., Nishimura, H., Matsuki, H., Osaka, F., & Kasuga, H. (1988). Urinary hydroxyproline to creatinine ratio as a biological effect marker for exposure to NO2 and tobacco smoke. *Atmospheric Environment, 22*(10), 2195-2203.

Yan, X., Komarneni, S., & Yan, Z. (2013). CO_2 adsorption on Santa Barbara Amorphous-15 (SBA-15) and amine-modified Santa Barbara Amorphous-15 (SBA-15) with and without controlled microporosity. *Journal of Colloid and Interface Science*, *390*(1), 217–224. doi:10.1016/j.jcis.2012.09.038 PMID:23084869

Yaumi, A. L., Abu Bakar, M. Z., & Hameed, B. H. (2017). Recent advances in functionalized composite solid materials for carbon dioxide capture. *Energy*, *124*, 461–480. doi:10.1016/j.energy.2017.02.053

Yeldan, E., Taşci, K., Voyvoda, E., & Özsan, M. E. (2013). Turkey on her way out of middle-income growth trap. Report for the Turkish Enterprise and Business Confederation.

Yi, H., & Feiock, R. C. (2014). Renewable energy politics: Policy typologies, policy tools, and state deployment of renewables. *Policy Studies Journal: the Journal of the Policy Studies Organization*, *42*(3), 391–415.

Yu, L., Batlle, F., & Carrera, I. (2011). Variations of waste unit weight during mechanical and degradation processes at landfills. *Waste Management & Research*, *29*(12), 1303–1315. doi:10.1177/0734242X10394912 PMID:21339240

Yumino, S., Uchida, T., Sasaki, K., Kobayashi, H., & Mochida, A. (2015). Total assessment for various environmentally conscious techniques from three perspectives: Mitigation of global warming, mitigation of UHIs, and adaptation to urban warming. *Sustainable Cities and Society*, *19*, 236–249. doi:10.1016/j.scs.2015.05.010

Zamfir, A., Colesca, S. E., & Corbos, R. A. (2016). Public policies to support the development of renewable energy in Romania: A review. *Renewable & Sustainable Energy Reviews*, *58*, 87–106.

Zhang, L., & Zheng, X. (2009). Budget structure and pollution control: A cross-country analysis and implications for China. *China & World Economy*, *17*(4), 88–103.

Zhang, M., Li, B., & Xiong, Z. Q. (2016). Effects of organic fertilizer on net global warming potential under an intensively managed vegetable field in southeastern China: A three-year field study. *Atmospheric Environment*, *145*, 92–103. doi:10.1016/j.atmosenv.2016.09.024

Zhang, S. X., Mab, T. Y., Renc, T. Z., & Yuan, Z. Y. (2011). Zirconium ribonucleotide surface-functionalized mesoporous SBA-15 materials with high capacity of CO2 capture. *Chemical Engineering Journal*, *171*(1), 368–372. doi:10.1016/j.cej.2011.04.021

Zhang, X., Pei, W., Deng, W., Du, Y., Qi, Z., & Dong, Z. (2015). Emerging smart grid technology for mitigating global warming. *International Journal of Energy Research*, *39*(13), 1742–1756.

Zhang, X., Qin, H., Zheng, X., & Wu, W. (2013). Development of efficient amine-modified mesoporous silica SBA-15 for CO_2 capture. *Materials Research Bulletin*, *48*(10), 3981–3986. doi:10.1016/j.materresbull.2013.06.011

Zhang, Y., Chu, C., Li, T., Xu, S., Liu, L., & Ju, M. (2017). A water quality management strategy for regionally protected water through health risk assessment and spatial distribution of heavy metal pollution in 3 marine reserves. *The Science of the Total Environment*, *599*, 721–731. doi:10.1016/j.scitotenv.2017.04.232 PMID:28499221

Zhao, D., Feng, J., Huo, Q., Melosh, N., Fredrickson, G. H., Chmelka, B. F., & Stucky, G. D. (1998). Triblock copolymer syntheses of mesoporous silica with periodic 50 to 300 angstrom pores. *Science*, *279*(5350), 548–552. doi:10.1126/science.279.5350.548 PMID:9438845

Zhao, X., Leiserowitz, A. A., Maibach, E. W., & Roser-Renouf, C. (2011). Attention to science/environment news positively predicts and attention to political news negatively predicts global warming risk perceptions and policy support. *Journal of Communication*, *61*(4), 713–731. doi:10.1111/j.1460-2466.2011.01563.x

Zheng, D., & Shi, M. (2017). Multiple environmental policies and pollution haven hypothesis: Evidence from China's polluting industries. *Journal of Cleaner Production*, *141*, 295–304.

Zhihui, H., Donghui, Z., & Jixiao, W. (2011). Direct synthesis of amine-functionalized mesoporous silica for CO_2 adsorption. *Chinese Journal of Chemical Engineering*, *19*(3), 386–390. doi:10.1016/S1004-9541(09)60225-1

Zhou, J., Zhao, H., Li, J., Zhu, Y., Hu, J., Liu, H., & Hu, Y. (2013). CO_2 capture on micro/mesoporous composites of (zeolite A)/(MCM-41) with Ca^{+2} located: Computer simulation and experimental studies. *Solid State Sciences*, *24*, 107–114. doi:10.1016/j.solidstatesciences.2013.07.008

Zierler, K. (1981). A critique of compartmental analysis. *Annual Review of Biophysics and Bioengineering*, *10*(1), 531–562. doi:10.1146/annurev.bb.10.060181.002531 PMID:7259129

About the Contributors

Bhola Ram Gurjar is a Professor in Civil (Environmental) Engineering, Dean of Resources & Alumni Affairs and Head of Centre for Transportation Systems (CTRANS) at India's premier technological institution, Indian Institute of Technology – IIT Roorkee. He holds a PhD in the area of Environmental Risk Analysis from India's premier technological institution I.I.T. Delhi followed by Postdoctoral research at the Max Planck Institute in Mainz (Germany). He has about 25 years' progressive professional experience in industry, teaching, training, research, and consultancy. Professor Gurjar has (co)authored / (co)edited nine books and 140 publications in total. He has received several awards and fellowships including the prestigious Advanced Postdoctoral Research Fellowship of the Max Planck Society (Germany) (2002-2005) and UKIERI Grant to visit Univ. of Surrey (U.K.) (Dec. 2012-Jan. 2013). Prof. Gurjar is among the leading academics and researchers who have worked extensively in the area of environmental science and engineering specially focused on air and water pollution, and environmental quality and health risk assessment, which is reflected in his several highly cited research papers published on these themes.

* * *

Ankur Chowdhury is presently pursuing PhD in the Department of Civil Engineering in Jaypee University of Information Technology, Waknaghat, Solan, Himachal Pradesh, India.

Rajiv Ganguly completed his Bachelor of Engineering in Civil Engineering from Nagpur University, India and his Master of Civil Engineering specialising in Environmental Engineering from Jadavpur University, India. Thereafter, he was a research student at Indian Institute of Technology-Guwahati for almost two years. He has done previous work on CNG and its viability as an alternative fuel to diesel and gasoline. He completed his PhD from the Department of Civil, Structural and

Environmental Engineering at Trinity College Dublin, Ireland working within the project 'Environment Transport Interface', responsible for evaluating various atmospheric dispersion models and their applicability to local environmental amendments. He also worked as a Postdoctoral Research Scientist at the School of Public Health in University of Michigan. Presently, he is working as an Associate Professor in the Department of Civil Engineering in Jaypee University of Information Technology, Waknaghat, Solan, Himachal Pradesh, India.

Ashok Gupta has 28 years of teaching experience and is presently Professor and Head of Civil Engineering Department, Jaypee University of Information Technology. He completed his Ph.D in Civil Engineering from I.I.T. Delhi and has been involved in teaching, research and involved in various administrative positions in the University.

Rajmal Jat received his M.Tech in Environmental Engineering and Management from Indian Institute of Technology, Kanpur, India in 2010. He was hired as the research engineer in Fluidyn Transoft Pvt. Ltd, in September 2011 and worked for one year. Afterward, during July 2011 to December 2016, he was engaged as the faculty member in Civil Engineering Department of Sir Padampat Singhania University, Udaipur, India. Since January 2017, he is associated as the PhD research scholar in the department of civil engineering, Indian Institute of Technology Roorkee, India.

Meenakshi Jatayan completed B.E. degree in Biotechnology from Maharshi Dayanand University, Rohtak, Haryana in 2009 and M.E. degree in Environmental Engineering from PEC University of Technology, Chandigarh, India in 2011. She is pursuing PhD in Environmental Engineering from PEC University of Technology, Chandigarh.

Kijpokin Kasemsap received his BEng degree in Mechanical Engineering from King Mongkut's University of Technology, Thonburi, his MBA degree from Ramkhamhaeng University, and his DBA degree in Human Resource Management from Suan Sunandha Rajabhat University. Dr. Kasemsap is a Special Lecturer in the Faculty of Management Sciences, Suan Sunandha Rajabhat University, based in Bangkok, Thailand. Dr. Kasemsap is a Member of the International Economics Development and Research Center (IEDRC), the International Foundation for Research and Development (IFRD), and the International Innovative Scientific and Research Organization (IISRO). Dr. Kasemsap also serves on the International Advisory Committee (IAC) for the International Association of Academicians and Researchers (INAAR). Dr. Kasemsap is the sole author of over 250 peer-reviewed international publications and book chapters on business, education, and information technology. Dr. Kasemsap is included in the TOP 100 Professionals–2016 and in the 10th edition of 2000 Outstanding Intellectuals of the 21st Century by the International Biographical Centre, Cambridge, England.

Shakti Kumar is Associate professor with Civil Engineering Department (Environmental Engineering Section), PEC University of Technology, Chandigarh, India.

Veerendra Sahu completed his B.tech in mechanical engineering from Uttar Pradesh technical university in 2012. In 2015 he received his M.tech in Environmental science and engineering from Indian Institute of Technology (ISM) Dhanbad and currently enrolled as PhD scholar in the department of civil engineering, Indian Institute of Technology Roorkee, India.

Index

www.ingramcontent.com/pod-product-compliance
Lightning Source LLC
LaVergne TN
LVHW081037190926
848465LV00005B/122